Studies in Systems, Decision and

Volume 148

Series editor

Janusz Kacprzyk, Polish Academy of Sciences, Warsaw, Poland
e-mail: kacprzyk@ibspan.waw.pl

The series "Studies in Systems, Decision and Control" (SSDC) covers both new developments and advances, as well as the state of the art, in the various areas of broadly perceived systems, decision making and control- quickly, up to date and with a high quality. The intent is to cover the theory, applications, and perspectives on the state of the art and future developments relevant to systems, decision making, control, complex processes and related areas, as embedded in the fields of engineering, computer science, physics, economics, social and life sciences, as well as the paradigms and methodologies behind them. The series contains monographs, textbooks, lecture notes and edited volumes in systems, decision making and control spanning the areas of Cyber-Physical Systems, Autonomous Systems, Sensor Networks, Control Systems, Energy Systems, Automotive Systems, Biological Systems, Vehicular Networking and Connected Vehicles, Aerospace Systems, Automation, Manufacturing, Smart Grids, Nonlinear Systems, Power Systems, Robotics, Social Systems, Economic Systems and other. Of particular value to both the contributors and the readership are the short publication timeframe and the world-wide distribution and exposure which enable both a wide and rapid dissemination of research output.

More information about this series at http://www.springer.com/series/13304

Anthony Tri Tran C. · Quang Ha

A Quadratic Constraint Approach to Model Predictive Control of Interconnected Systems

 Springer

Anthony Tri Tran C.
Cambridge CARES
Singapore
Singapore

Quang Ha
University of Technology Sydney
Sydney, NSW
Australia

ISSN 2198-4182   ISSN 2198-4190   (electronic)
Studies in Systems, Decision and Control
ISBN 978-981-13-4142-7   ISBN 978-981-10-8409-6   (eBook)
https://doi.org/10.1007/978-981-10-8409-6

Printed on acid-free paper

This Springer imprint is published by Springer Nature
The registered company is Springer Nature Singapore Pte Ltd.
The registered company address is: 152 Beach Road, #21-01/04 Gateway East, Singapore 189721,
Singapore

Preface

The attraction of achieving higher efficiency and reliability for industrial plants and networked systems has created new research opportunities in the control and optimisation field. Among different design methods, the model predictive control (MPC) strategies, first developed for the petroleum refining industry, have proved to be effective in many applications. Originally found a widespread use in the stand-alone sites, the non-centralised adaption such as distributed and decentralised MPCs have been progressing towards more heterogeneous architectures that are able to cope with system complexities and variations in application domains.

This book presents a stabilising method for the control of interconnected systems having mixed connection configurations with distributed and decentralised model predictive control schemes. The novel notions of asymptotically positive realness constraint (APRC) and quadratic dissipativity constraint (QDC) are introduced as a fundamentally constituent part of this book. In both constraints, the function of inputs and outputs in the form of a supply rate, or a 'supply power', is quadratic. From the communication and information perspective, the quadratic constraint packs two pieces of information, the control and state vectors, into one variable, before carrying to different locations, and then unpacks them for use with the local control algorithm. The employment of quadratic constraints in two distinct approaches, segregation from and integration into the control algorithms, for the constrained stabilisation of interconnected systems is another contribution of this book.

Solutions for linear systems are given in distributed and decentralised strategies whereby the communication between subsystems is either fully connected, partially connected or totally disconnected. The interconnected systems and their distributed computerised-control platforms are considered within the realm of a cyber-physical system consisting of the physical connections between subsystems and the communication links between local computing processors. Within the auspice of the integrated construction method, the distributed and decentralised MPC strategies deal with the communication links from the cyber-connection side—the subsystems are wholly or partially connected in a distributed MPC scheme while being totally disconnected in a decentralised MPC.

By having the control inputs entirely or partially decoupled between subsystems, and no additional constraints imposed on the interactive variables, rather than the coupling constraint itself, the proposed approaches outreach various types of networked systems and applications. The effects of coupling delays and device networks are also resolved in part of the development. For parallelised connections that emulate parallel redundant structures and have unknown splitting ratios, a fully decentralised control strategy is developed as an alternative to the hybrid system approach. For the semi-automatic control systems, involved with both closed-loop and human-in-the-loop regulatory controls, the stability-guaranteed method of decentralised stabilising agents, that are interoperable with different control algorithms, is germinated and implemented for each single subsystem.

For nonlinear input-affine systems, the extended output vector including the vector field and the state vector are introduced such that the dissipativity criterion can be rendered in linear matrix inequalities. The compound vector can be viewed as manifest variables in the beviourial framework for dynamical systems. From the perspective of the dissipative system theory, both the storage function and supply rate with the extended output vectors are parameterised to avoid any conservativeness that may incur to the stabilisation of nonlinear systems.

In this book, MPC is formulated with state-space models having a standardised cost function. The stability constraint here is a constraint imposed on the current-time control vector, independently to the MPC objective function. For interconnected systems, the terminal constraint computations are formidable when dealing with subsystems having dissimilar dynamics whose settling times are heterogeneous. The quadratic constraint approach resolves this difficulty by having a constraint on the current-time control vector. The state constraint and recursive feasibility are, nevertheless, not included in this book.

An extension to new applications with the Internet of Things (IoT) is also presented with some dependable control schemes in which multiple controllers and sensors are cross-connected via the IoT communication network to ensure the duty-standby architecture for achieving quantitatively higher reliability of cyber-physical systems.

A broad range of applications in the process and manufacturing industries, networked robotics, networked control systems and network-centric systems such as power systems, telecommunication networks and chemical processes will benefit from the approaches in this book. Illustrative examples of networked interconnected systems are provided with numerical simulations in MATLAB environment. Specifically, a power system having four control areas, a dependable controller for cyber-physical systems and some other numerical examples are implemented with the distributed and decentralised MPC strategies employing the quadratic constraint approach to demonstrate the theoretical appraisals.

The developments are presented in seven chapters. This book starts with an introduction to the quadratic constraint in the time domain with a different perspective, as stated in Chap. 1. Here, the differences between this closed-loop perspective on the dissipation-based constraint and the other open-loop dissipative system approaches in the well-known interconnection stability conditions with passivity and small-gain theorems will be highlighted. A brief review on the MPC applications and the stabilising methods for the previously developed distributed MPC strategies is also given in the first chapter. Chapter 2 is dedicated to the quadratic constraints and their applications to the decentralised MPC of interconnected systems as the enforced attractivity constraints. In the next chapter, the attractivity conditions for the complex interconnected systems that have parallelised connections with unknown splitting ratios are presented, Chap. 3. An alternative constructive method of stabilising agents with the QDC is then delineated following in Chap. 4. Chapter 5 outlines a deterministic approach to the data lost processes with the presented dissipation-based quadratic constraints. A virtual perturbed cooperative-state feedback (PSF) strategy will be presented in the second part of this chapter. The available communication network in a cyber-physical system is capitalised on for improving the control performance with the PSF strategy. The developments for interconnected systems having a coupling delay element with the accumulative quadratic constraints are subsequently provided in Chap. 6. Chapter 7 is dedicated to the QDC application to the dependable control systems.

The general dissipativity constraint (GDC) method for the control design and synthesis of multi-variable systems in the discrete-time domain is presented in Appendix A. APRC and QDC with quadratic supply functions are the two special cases of the GDC. The dissipation-based constraints with a general supply function and the stability with a relaxed non-monotonic Lyapunov function and the input-to-power-and-state stabilisability (IpSS) are presented in this appendix. The GDC method for stabilising the interconnected systems with distributed, decentralised and dependable control architectures is well suited to the modern cyber-physical systems incorporating scalable and flexible communication networks. With emerging technologies in the Internet of Things (IoT) and cloud computing, the new architecture and algorithms will provide the tractability for implementations in a connected and 'smart' environment, yet help achieve the required reliability and continuity of the operational systems. The well-known MPC algorithms that employ plant models in the future state prediction for computing the control moves with convex optimisations have been found agile for deploying with cyber-physical systems.

During the course of preparation of this monograph, there were a series of invaluable discussions with Profs. Jan Maciejowski, Hung T. Nguyen and Tuan D. Hoang, to whom the authors are much indebted. In particular, the first author would like to gratefully acknowledge support obtained from the Singapore National Research Foundation (NRF) under its Campus for Research Excellence And Technological Enterprise (CREATE) programme and the Cambridge Centre for Advanced Research and Education in Singapore (Cambridge CARES), C4T project. Support received from various internal grant schemes at the Faculty of Engineering and Information Technology and the University of Technology Sydney, Australia, is also acknowledged.

Sydney, Australia   Anthony Tri Tran C.
November 2017   Quang Ha

Contents

List of Figures

Chapter 1
Introduction

1.1 General

Automatic and semi-automatic control of large-scale interconnected systems, also known as network systems [10, 101], remains a challenging problem due to the increasing interactions between multi-variable subsystems connected in parallel, parallelised (Chap. 3), serial and recirculation. In a broad sense, a network system can be modelled as a large set of interconnected nodes, in which a node is a fundamental unit with specific contents [81]. The network system notion is extended over many application domains such as, but not limited to, chemical and petrochemical processes, power systems, telecommunication systems, transportation systems, supply chain systems, networked robotics and biological systems.

Figure 1.1 depicts a conceptual block diagram of such network systems with parallelised, serial and recirculated connections. In this book, we consider an interconnected system with a distributed computer system and a sensor/device network, as shown in Fig. 1.2a, b, that form a cyber-physical system (Chap. 5). The term cyber-physical system came into effect recently to describe the integration of control, computation, networking, physical processes and human–machine–machine interactions. In other words, cyber-physical systems (CPS) are "smart systems that include co-engineered interacting networks of physical and computational components", as defined by the National Institute of Standards and Technology (NIST). Despite the generic nature of the content covered in this book, some case studies and numerical examples provided herein are mainly for chemical process systems and power systems.

Feedback control of large-scale systems is a classical topic in the control theory, see, e.g. [1–3, 86, 100, 131, 140]. The coordination methods for the control of interconnected systems and various decomposition techniques have been used for many years, see, e.g. [142]. It is not, however, trivial to apply those approaches directly to the modern control problems with mathematical programming and flexible features of on-the-fly adjustments in a real-time computing environment. As a matter of course, the distributed and decentralised model predictive control of large-scale

© Springer Nature Singapore Pte Ltd. 2018
A. Tri Tran C. and Q. Ha, *A Quadratic Constraint Approach to Model Predictive Control of Interconnected Systems*, Studies in Systems, Decision and Control 148, https://doi.org/10.1007/978-981-10-8409-6_1

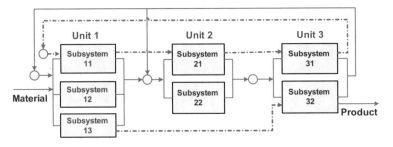

Fig. 1.1 Mixed connection configuration of an interconnected system

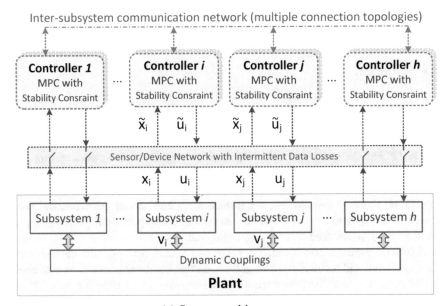

(a) System architecture

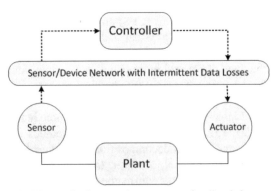

(b) Networked control system - a feedback loop

Fig. 1.2 An interconnected system under the cyber-physical system auspice

interconnected systems [89, 125] have only been developed during the last two decades. The field has been progressing towards a heterogeneous architecture and outreaching different application domains.

The MPC algorithms cannot assure the closed-loop system stability in a cyber-physical environment without having an extra measure. To this end, the stability-guaranteed problem of MPC with state and control constraints has long been a research topic that draws keen interests from both academia and industry, see, e.g. [4, 5, 13, 19, 20, 23, 74, 75, 80, 88, 90, 93, 115, 118, 124, 128]. A thorough review on the stability and optimality of different MPC formulations and stability-guaranteed methods can be found in [93, 117]. The constrained closed-loop system with MPC is not guaranteed stable if the predictive horizon N is not sufficiently long [88], and further, the MPC optimisation may not be recursively feasible if the initial state vector does not belong to an appropriate admissible set [93]. The recursive feasibility may imply system stability regardless of the length of the predictive horizon N if an additional constraint is imposed onto the MPC optimisation. Among the developed stability constraints, the terminal state constraint associated with the maximal output admissible set [42, 73] is well known in the MPC literature. However, the terminal constraint approach may provide a conservative feasible region for distributed MPCs, especially for heterogeneous, dynamically coupled systems, see, e.g. [24, 121]. The quadratic constraint approach offers a non-conservative alternative for the stability assurance of decentralised and distributed MPCs.

The quadratic constraint in this book has been developed on the ground of the dissipative system theory [172], which is attractive to the control design problem for interconnected systems owing to its input and output usage and related properties. Although the theory has been applied to the stability analysis and control design of linear and nonlinear systems since the 1970s, see, e.g. [17], the practical implementations have not been found widespread used yet. Inspired by the (Q, S, R)-dissipativity in [59, 133, 134, 159, 172], the quadratic supply-rate function has been extensively employed in this work. Our work is focused on the open-loop approach (or the closed-loop perspective), as we believe it is suitable for the decentralised model predictive control schemes. The interconnection stability conditions with the two open-loop dissipative systems are not applicable in this open-loop approach, as a result, but the conditions on the closed-loop system are derived instead. Under this open-loop realm, the quadratic supply rate can be developed into a stability constraint for the MPC, which is a constraint on the current-time control vector. The coefficient matrices of the stability constraint can be updated online, or simply pre-computed off-line, for guaranteeing the closed-loop system stability. The notions of asymptotically positive realness constraint (APRC) and quadratic dissipativity constraint (QDC) are introduced here in this context.

The simplest form of APRC that employs a real-valued quadratic supply function with respect to the state and input pair, $\xi(x_i(k), u_i(k))$, is expressed as a constraint in $u_i(k)$, when $x_i(k)$ and $\xi(x_i(k-1), u_i(k-1))$ are known. For example, the key conditions with the APRC consist of the dissipation inequality with a storage function, as in the dissipative systems theory [172], of the form

$$V(x_{k+1}) - \tau V(x_k) \leqslant -(x_k^T \ u_k^T)N(x_k^T \ u_k^T)^T, \quad V(x_k) \geqslant 0,$$

where $(x_k^T \ u_k^T)^T$ is the stacking vector comprising x_k and u_k, and a dissipation-based inequality of the form

$$0 \geqslant (x_k^T \ u_k^T) N (x_k^T \ u_k^T)^T \geqslant \gamma \, (x_{k-1}^T \ u_{k-1}^T) N (x_{k-1}^T \ u_{k-1}^T)^T \ \forall k \in (0, k_s],$$

$$\text{and } (x_k^T \ u_k^T) N (x_k^T \ u_k^T)^T \geqslant 0 \ \forall k > k_s > 0,$$

where $N := \begin{pmatrix} Q & S \\ S^T & R \end{pmatrix}$, $\gamma \in (0, 1)$ and $\tau \in (0, 1)$.

The key conditions with the QDC also consist of the dissipation inequality with a storage function and a dissipation-based inequality, as follows:

$$V(x_{k+1}) - \tau V(x_k) \leqslant -(x_k^T \ u_k^T) N (x_k^T \ u_k^T)^T, \quad V(x_k) \geqslant 0,$$

$$0 \geqslant (x_k^T \ u_k^T) N (x_k^T \ u_k^T)^T \geqslant \gamma \, (x_{k-1}^T \ u_{k-1}^T) N (x_{k-1}^T \ u_{k-1}^T)^T \ \forall k > 0.$$

The differences between the APRC and the QDC are in the second inequalities whose fulfilment may not incur at every time steps in $\mathbb{Z}$ or only within a finite-time interval. We will prove that the attractivity can be obtained from the above APRC and QDC inequalities for $V(x_k) = x_k^T P x_k$, where P is full row rank. Furthermore, a generalised form of the dissipation-based inequality with a $\mathscr{K}\mathscr{L}$−bounded supply function (see Sect. 1.3.2) will also be introduced, the general dissipativity constraint (GDC). In the GDC method, we often deal with the non-negative $\triangle V(x_k)$, i.e. $V(x_k) - V(x_{k-1}) \geqslant 0$ (and decreasing, not necessarily monotonically), but not with $\triangle V(x_k) \leqslant 0$, as in Lyapunov's method.

The APRC and QDC are subsequently recast and incorporated into the online optimisation problem of a local MPC as an enforced attractivity constraint, or employed in the algorithm of stabilising agent independently to the associated control algorithms. These quadratic constraint approaches are non-conservative for the decentralised control problem by virtue of its non-monotonic storage function that plays a similar role as a relaxed non-monotonic Lyapunov function whose non-conservativeness has been pointed out in [96].

Methodologically, the online-centric stabilisation of interconnected systems with decentralised model predictive controllers is the main theme of this book. Convex programming numerical methods with their elegant solutions [15, 103] are used throughout the numerical examples. By the same token, the linear matrix inequality (LMI) is extensively employed in the derivations for dissipative criteria and stability conditions for interconnected systems.

For semi-automatic control with quadratic constraints in Chap. 4, the control system is associated with a stabilising agent that is interoperable with different control algorithms. This segregated stabilising method can be applied to the stability assurance problem for the local and remote operations of a cyber-physical system. The stabilising agent relies on the quadratic constraint trajectory to initiate corrective actions to stabilise the operational system. Both intermittent data losses and coupling delays are inclusive and resolved deterministically by extending the quadratic constraint and dissipative conditions to recover the convergence property.

We formulate the redundant processes as continuously parallelised, or parallel splitting, systems in Chap. 3 as an alternative to the hybrid system with Boolean variables. As a result, the global system is effectively stabilised in different operational scenarios, including concurrent and duty-standby modes. The parallelism characteristic comes from the splitting ratios of parallelised plants, which are unknown and usually time dependent. The concept and mechanism of self-recovery control and dependable control systems using the Internet of Things (IoT) are introduced for the building of a reliable operational technology (OT) (computerised control) system, within a connected environment of cyber-physical systems in Chap. 7.

The implementation aspect of the quadratic constraint method, but not the mathematical theory, is emphasised in this book. The remaining of this chapter is reserved for introducing the quadratic constraints in the time domain with a practical perspective. A brief review of passivity and small-gain theorems is provided at the end of the chapter to simply show that they are not applicable to the quadratic constraint approaches in this book.

1.2 Quadratic Constraint—a Time-Domain Perspective

A practical and time-domain perspective of the quadratic constraint in a control problem is presented in this section.

We first discuss the meaning of the passivity term, $-y^T u$, or $-y \times u$ for systems with single variables, in a feedback control system shown in Fig. 1.3, and various trajectories of $-y^T u > 0$, or $-y^T u \geqslant 0$, cases, as shown in Fig. 1.4, within the time domain.

From the theoretical perspective, the passivity term represents the phase property of an output and input pair, see, e.g. [139]. It is well known that the phase and gain properties are important for the control system design in the frequency domain with Bode and Nyquist plots in the control literature [3, 38]. However, the motion of the passivity term,

$$-y(t)^T u(t),$$

or of the quadratic supply rate of the form

$$-\alpha\, y(t)^T y(t) - \beta\, y(t)^T u(t) + \gamma\, u(t)^T u(t),$$

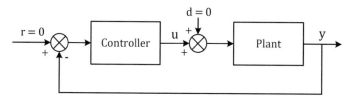

Fig. 1.3 A feedback control system

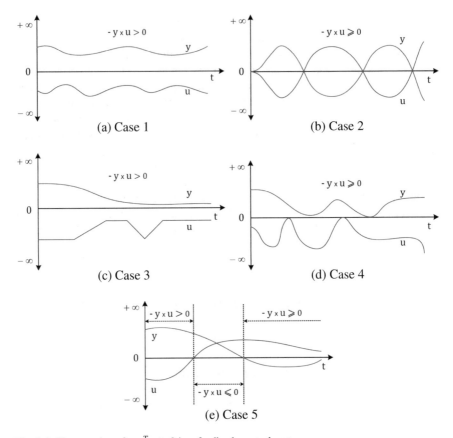

Fig. 1.4 The meaning of $-y^T u \geqslant 0$ in a feedback control system

in the time domain does not necessarily convey a stability or convergence message. The five cases of $y(t)$ and $u(t)$ trajectories shown in Fig. 1.4 illustrate the typical examples of $-y^T u \geqslant 0$, which suggest irrelevant attractiveness of the related controlled system. In what follows, the time index $k \in \mathbb{Z}^+$ is used for discrete-time systems.

1.2.1 Positive Supply Power

The discussion starts with a special case of relationship between the input and output of an *absolute energy-passive* (AP) signal system, such as the input and output of a feedback control system. We will show that the concept of *energy-passivity* can be adopted to illustrate the connotation of dissipating the *abstracted energy* around the steady states, and the *energy-dissipativity* may imply a converging output and input pair or an attractive controlled system.

Fig. 1.5 An absolute energy-passive (AP) motion

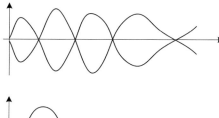

Fig. 1.6 A perfect one-overshoot stable (POS) motion

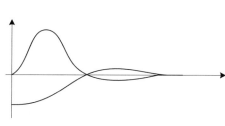

An input and output signal system is said to be absolute energy-passive (AP) if and only if both of its input and output trajectories, $u_{(k)}$ and $y_{(k)}$, respectively, always pass through their corresponding steady states (or set point) at the same time instants.

For clarity, the output and input steady states can be assumed to be zero ($y_{ss} \equiv 0$, $u_{ss} \equiv 0$), i.e. normalised to zero steady states. An AP motion is not necessarily attractive or convergent. Figure 1.5 depicts a non-attractive AP motion. For signal systems that are AP, the following discrete-time inequality holds true for all $k \geqslant 0$:

$$p_{(k)} := (y_{(k)} - y_{ss})^T (u_{ss} - u_{(k)}) \geqslant 0. \tag{1.1}$$

The inequality (1.1) is called *instant absolute energy-passive inequality* (iAPi) herein. An AP signal system is nothing else than a special case of passive systems and a subclass of zero input–output phase shift systems.

The absolute energy-passivity is apparently difficult to be obtained from a feedback control system. From the feedback control and circuit theory point of view, we are familiar with the overshoot of under-damped control systems. Therefore, the so-called *perfect one-overshoot stable* (POS) motion is introduced next.

Figure 1.6 depicts a typical POS motion. While AP motions are not necessarily attractiveness or convergence, the POS motion is a special case of stabilised and well-performed control systems.

The critically damped controlled system is perfectly one-overshoot stable, or POS, if and only if it is AP and the output and input motions are attracted to their steady states after exactly one overshoot.

Similar to AP signal systems, it is not easy to obtain a POS behaviour from a control design problem. We usually have a stabilised controlled system that also has an overshoot, but does not have the AP property; i.e., the input and output trajectories do not cross their steady states at the same time instants, as shown in Fig. 1.7. This is the time response of a typical critically damped system in the feedback control literature. One can easily verify that the sum of $p_{(k)}$ over the time intervals of T_1 and

T_3 is greater than that over T_2 for some stabilised controlled systems. Further, the critically damped system will have the AP property when $T_2 \rightarrow 0$. We now link a stabilised controlled system with the energy-passivity defined below.

For an energy-passive and stabilised controlled system, the accumulative sum of $p_{(k)}$ over time is always non-negative, i.e.

$$\sum_{\kappa=0}^{k} (y_{(\kappa)} - y_{ss})^T (u_{ss} - u_{(\kappa)}) \geqslant 0 \; \forall k \geqslant 0. \tag{1.2}$$

Alternatively, it can be said that, for an energy-passive and stabilised controlled system, the sum $s_{(k)}$ of $p_{(k)}$ where $p_{(k)} \geqslant 0$, denoted as $s_{(k)}^+$, is greater than or equal to the absolute value of the sum $s_{(k)}$ of $p_{(k)}$ where $p_{(k)} < 0$, denoted as $s_{(k)}^-$, i.e.

$$s_{(k)}^+ \geqslant -s_{(k)}^-. \tag{1.3}$$

An example for the trajectories of a more general case of an energy-passive and stabilised controlled system is shown in Fig. 1.8. In this figure, the sum of $p_{(k)}$ from odd intervals T_1, T_3, T_5, ... ($= s_{(k)}^+$) is greater than the sum of $p_{(k)}$ from even intervals T_2, T_4, T_6, ... ($= s_{(k)}^-$). The even time intervals represent the phase differences between the input and output trajectories. It can be seen that this type of trajectories is achievable from a practical control design problem and that if $T_{\text{even}} \rightarrow 0$, the controlled system will become AP or POS, which is impractical for a control design problem.

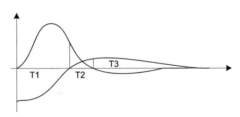

Fig. 1.7 POS and energy-passive, but not AP

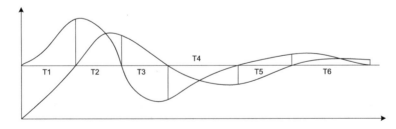

Fig. 1.8 Energy-passive and stabilised controlled system—an example of input and output trajectories

1.2.2 Energy-Dissipative Motion

The *energy-dissipativity* is a generalisation of the energy-passivity and different to the dissipativity in the control literature. Under the energy-dissipative perspective, $p_{(k)} = (y_{(k)} - y_{ss})^T (u_{ss} - u_{(k)})$ is considered as the supply power, while $s_{(k)}^+$ is the consumed energy, and $s_{(k)}^-$ is the generated energy. When (1.3) holds, we say that the signal system is energy-dissipative; i.e., the *abstracted energy* is dissipated away. The two components of the supply power, *consumed energy* and *generated energy*, have not been defined in the dissipative systems theory. However, the energy storage can also be made available with the energy-dissipativity, as follows: recalling the energy conservation rule in nature, we can state that the availability of a reserved energy $\mathscr{E}$, and its non-negative motion, $V_{(k)} \geqslant 0$, will provide a larger margin for the inequality (1.3) of an energy-passive and stabilised system to be fulfilled,

$$s_{(k)}^+ + s_{(k)}^- \geqslant \mathscr{E}_{(k)}, \tag{1.4}$$

instead of $s_{(k)}^+ + s_{(k)}^- \geqslant 0$, and so will the abstracted energy-dissipation.

If we draw the trajectories of $|p_{(k)}|$ for the supply power of Fig. 1.8, as shown in Fig. 1.9, the accumulative sum $s_{(k)}$ of $|p_{(k)}|$ is represented by the cross sections of areas under the curve. This figure apparently explains why $s_{(k)}$ can be viewed as the abstracted energy of the supply power $p_{(k)}$, and the area width and height are metrics for this abstracted energy quantity. The width has the phase-related information, while the height carries the amplitude-related information. We will show that these two metrics will affect the controlled system behaviours when being adjusted.

Talking about the phase- and amplitude-related quantities, we are introducing the two components of the *energy-dissipativity*. The energy-dissipativity contains the energy-passive components with phase-related property and the amplitude components with gain-related property. The quadratic supply power of the following form

$$p_{(k)} := y_{s(k)}^T Q y_{s(k)} + y_{s(k)}^T S u_{s(k)} + u_{s(k)}^T R u_{s(k)},$$

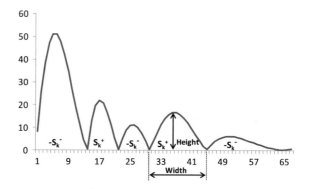

Fig. 1.9 Abstracted energy from an energy-dissipative behaviour

where $y_{s(k)} := y_{(k)} - y_{ss}$ and $u_{s(k)} := u_{(k)} - u_{ss}$, in which y_{ss} and u_{ss} are the corresponding steady states, thus represents a proper form of the supply power in an energy-dissipative system. When both the consumed energy and generated energy converge to a steady-state point, we have an attractive motion. The following hypotheses consolidate the discussion:

An energy-dissipative and stabilised controlled system possesses a non-negative accumulation of the supply power, i.e.

$$\sum_{\kappa=0}^{k} P_{(\kappa)} \geqslant 0 \ \forall k \geqslant 0. \tag{1.5}$$

An energy-dissipative controlled system possesses an energy reserve; i.e., the following energy-dissipative inequality is satisfied for all $k \geqslant 0$:

$$\sum_{\kappa=0}^{k} P_{(\kappa)} \geqslant \mathscr{E}_{(k)}. \tag{1.6}$$

With the introduction of the supply power, its accumulation as the supply energy, and the energy-dissipative concept, and also linking them to a stabilised control system, we will show that the supply power and energy-dissipative inequality (1.6) can be employed in a control design problem for both stand-alone and interconnected systems.

One may easily find that, an adjustment of the control signal $u_{(k)}$ will instantly change the amplitude quantity, but may not have an immediate effect on the phase quantity. Some forms of prediction should be applied if the adjustments aim to change the phase-related quantity, i.e. the energy-passivity. Along this line, the asymptotic property of a supply power and supply energy is probably well suited to a synthesising strategy with predictions for an energy-dissipativity-based method. Finally, it is apparently that the inequality (1.6) is a special case of the general dissipativity inequality in the discrete-time domain,

$$-\sum_{\kappa=0}^{k} P_{(\kappa)} \leqslant \beta_d \ \forall k \geqslant 0,$$

presented by Willems J. in [173] for dissipative systems without a storage function, which is called the "no-thrilled" definition for dissipative behaviours.

The quadratic constraint method with the GDC, including the APRC and QDC, for multi-variable control system design, especially with the MPC algorithms, has been geminated from the hypotheses in this section. The APRC or QDC will be employed in the MPC optimisation as an enforced attractability constraint, similar to the other enforced stability constraint approaches for MPC, see, e.g., [93]. The energy-dissipativity will be sufficiently obtained as a result of associating the energy-dissipative-based quadratic constraint with the MPC optimisation problem. Chapter 2 presents the attractability constraint for the MPC with the energy-dissipativity. Alternatively, the quadratic constraint can act as a *stabilising agent* independently to the control algorithm and will override the relevant control actions whenever the quadratic constraint is violated. This segregation of the control algorithm from the stabilising mechanism will enhance the system reliability simultaneously to the closed-loop system stability. Chapter 4 presents the stabilising agent with the quadratic constraint method.

In the next section, we illustrate the notion of energy-dissipative controlled system and energy-dissipativity hypothesis, in a design example with the proportional–integral–derivative (PID) controllers. This design procedure is not for a practical problem, but for the illustration purpose only.

1.2.3 Predictive PID Based on Energy-Dissipativity

A gain adjustment algorithm for the PID controller of a single-input-single-output (SISO) linear-time-invariant (LTI) open-loop stable system based on the energy-dissipativity is presented here for the purpose of illustration. The P, I and D gains are adjusted on the fly to ensure that the system energy-dissipativity will be obtained at every time step $k > 0$. Since this is only an illustrative algorithm, a simple Excel tabular sheet has been used for the predictive calculations and a second-order plus dead-time plant is studied.

A PID controller with high gains is initially implemented to produce a limit-cycle-like response. The online adjustment strategy is to change the PID gains so as the system time response approaches an energy-passive behaviour after some future moves. The algorithm iterates predictive values of $p(k)$ towards the end of a chosen horizon N. The horizon N must be greater than the time instant T_s when the output trajectory exceeds its set point for the first time. The PID gains are updated to increase T_s after every iteration, i.e. $T_{s(k)} > T_{s(k-1)}$, where $T_{s(k)}$ denotes T_s at the iteration k. The PID gains are initially reduced in small decrements based on two inequalities, $p(N) \geqslant 0$ and $T_{s(k)} > T_{s(k-1)}$. After some predictions, if reducing PID gains does not recover the iAPi condition beyond N, i.e. $p(N) \geqslant 0$, then start increasing them by some small increments. This process will stop when the output stays around the set point steadily.

The resultant time response is almost identical to a POS as depicted by Fig. 1.11a, b. Figure 1.10a shows the time response with large and fixed PID gains. Figure 1.11a indicates a significant improvement on the control performance when the above predictive PID adjustment is employed, compared to that in 1.10(a). The supply power trajectories $p(k)$ are given in Figs. 1.10b and 1.11b, respectively. The online PI gains are provided in Table 1.1 (D term is not used). The transfer function of the open-loop system used in this simulation is as follows:

$$G(s) = \frac{0.25}{s^2 + 0.5s + 0.25} e^{-7s}.$$

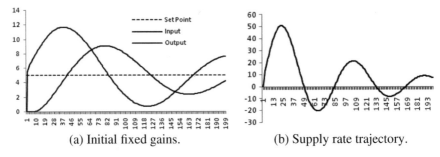

(a) Initial fixed gains.   (b) Supply rate trajectory.

Fig. 1.10 High-gain PID control

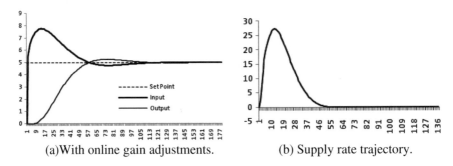

(a)With online gain adjustments.   (b) Supply rate trajectory.

Fig. 1.11 Illustrative predictive PID based on energy-dissipativity

Table 1.1 Predictive PI
controller gains

Step No.	K_p, K_I	Horizon
1	0.1, 0.5	50
6	0.05, 0.2	50
8	0.3, 0.1	52
11	0.02, 0.1	52
14	0.0075, 0.001	55
17	0.002, 0.2	59

   The coarse adjustments have also been simulated without looking at the constraints
on $\Delta u_{(k)}$. The resulting trends are given in Fig. 1.12. The corresponding online PID
gains are provided in Table 1.2 given next.
   The trajectory of $p(k)$ in this case is not very different to the previous fine adjustment, but the control trajectory is quite different, as shown in Fig. 1.12. According to
the time responses of the supply power $p(k)$ in the above figures, we can see that it
is crucial to maintain a continuous $p(k)$ trajectory for perceiving a realisable control
performance for continuous plants.
   This case study gave us some intuitive insights into the behaviours of input and
output trajectories against the online adjustment based on the energy-dissipativity.
We have shown that the supply powers can be adjusted in this example to achieve
the energy-dissipation and, thus, the desirable performance for the feedback control
system.

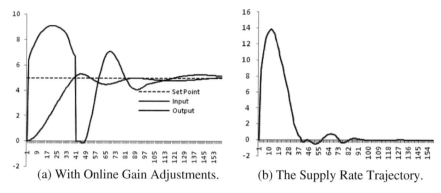

(a) With Online Gain Adjustments.   (b) The Supply Rate Trajectory.

Fig. 1.12 Illustrative predictive PID—coarse adjustments

Table 1.2 Predictive PI with coarse adjustments

Step No.	K_p, K_I	Overshoot
1	0.08, 1.2	6.55
5	0.06, 1	6.35
35	0.2, 3	5.65
38	0.1, 2	5.6
42	0.0001, 0.0001	5.2
47	0.2, 8	5.3
100	0.1, 1.2	0.3

1.3 General Dissipativity Constraint

The stability of a controlled passive system is usually studied with passivity theorems, see, e.g. [59, 139], which stipulate the finite gain stability criteria. The application of passivity theorems and Lyapunov stability to the analysis and design of control systems has been studied and is well established in the control theory [17, 166]. The sufficient condition for bounded-input-bounded-output (BIBO) stability in a small-gain theorem [26, 38, 180] is also well known with the property of interconnected signals in the extended space $\mathscr{L}_e$ of the normed linear subspace $\mathscr{L}$. The small-gain design has, however, been shown more conservative than the circle criterion and Popov criterion designs [54] for linear systems. It has also been shown that the "phase" information can be used in the control design to reduce the possible conservativeness from the small-gain theorems and quadratic Lyapunov functions [52], as reflected in the standard feedback control design with the strictly positive real compensators. Introduced by Sontag E.D. [144, 145], the input-to-state stability (ISS) has extended the use of classical Lyapunov's methods to systems with inputs, by "exploiting energy-like functions in order to assess the stability and robustness of a system with respect to internal and external perturbations" [6]. Unfortunately, the constructive methods for ISS have usually been developed around some small gains for a closed-loop system and thus may limit their applications.

Inspired by the energy-dissipativity hypothesis, we have developed the (quadratic) dissipation-based constraint method to render the *Input-to-power-and-State Stability* (IpSS) of controlled systems having internal and external perturbations. The IpSS is not ISS since the stability without disturbances is not the Lyapunov stability. The storage function (quasi-Lyapunov) in the IpSS is merely a relaxed non-monotonic Lyapunov function.

The dissipation-based inequality without a storage function, similar to the 'no-thrilled' dissipation inequality defined by Willems J. in [173], will be an integrated part of the presenting method, in which the supply rate is monotonically decreasing along the trajectories or $\mathscr{K}\mathscr{L}$−bounded. In addition to that, the supply function for nonlinear input-affine systems is extended with a compound output vector. The compound vector is a reflection of manifest variables under the auspice of the behavioural framework for dynamical systems [114]. We, however, restrain our developments within the state space in this work. In the dissipative systems theory, one usually refers to the input variable u in the supply rate of the dissipation inequality as either disturbance, reference, or control input, while the output variable y can be either measurement output, functional output, or simply state. The supply function is then denoted as $\xi(y, u)$. Here, we extend the output vector to include some functions of the state vector. Specifically, for $x^+ = f(x) + g(x)u + d$, the compound output will consist of the state vector x and the vector field f, i.e. $[x^T \ f^T]^T$. The supply function becomes then a real-valued function w.r.t $[x^T \ f^T]^T$ and the compound input vector $[u^T \ d^T]^T$. When the quadratic supply function with respect to the compound vectors is chosen, it is possible to derive LMI-based dissipative criteria for discrete-time nonlinear input-affine systems, compatible with other well-known LMI criteria for dissipative systems in the control literature, see, e.g. [40, 135, 138].

The input-to-state stability and its variants have been used in research work since the 1990s. The prevailing stability criteria based on ISS are Input-to-State Practical Stability (ISPS) [63], Output-to-State Stability (OSS) [64], Input-Output-to-State Stability (IOSS) [76], Input-to-State Dynamical Stability (ISDS) [47], integral Input-to-State Stability (iISS) [7], and their variants for continuous, hybrid, switched, time-varying and stochastic systems. The discrete-time version of the ISS has been presented in [18, 65, 67]. A recent review on ISS can be found in [25]. The non-conservative small-gain conditions have recently been discussed in [41]. The dissipation-based quadratic constraint presented here renders an extended ISS for nonlinear systems in the discrete-time domain—the IpSS with relaxed non-monotonic Lyapunov functions. The new stabilising method is suitable for the emerging large-scale cyber-physical systems with distributed and decentralised MPC of interconnected and network systems.

The control design and synthesis methods for dissipative systems and systems having $\mathscr{L}_2$ gain have been predominated by the passivity and small-gain theorems, in which the plant and controller are dissipative or strictly dissipative, or both have $\mathscr{L}_2$ gains. The Lyapunov stability of the equilibrium is achieved with those methods. In this work, we have the controlled motions satisfied a dissipation-based constraint to obtain the attractability with a relaxed non-monotonic Lyapunov function.

The idea of having a dissipation-based constraint on the controller in a control design problem has been inaugurated in recent years, such as those in [21, 71,

79, 120, 151, 153, 154, 165]. However, the presenting general dissipativity constraint (GDC) method is different to those previous works by newly introducing a dissipation-based constraint to the closed-loop system such that both dissipative and attractive conditions are obtained simultaneously. The open-loop system dissipativity is not a prerequisite condition in this GDC method. The storage function in this development is a relaxed non-monotonic Lyapunov function and thus less conservative than the monotonic Lyapunov functions [96] in the well-known passivity and small-gain theorems. Furthermore, the storage function is not only non-monotonic, but also piecewise continuous, or discontinuous, along the trajectories in general, whereas the Lyapunov function or ISS Lyapunov function is usually smooth or locally Lipschitz continuous. As a result, the GDC approach is less conservative for the decentralised MPC problem of heterogeneous interconnected systems whose subsystems are dynamically coupled with a complex coupling structure (connection topology) and networked controlled.

In this book, we adopt the notions of 'dynamical system' and 'motion' in [96], as follows:

A *dynamical system* is a four-tuple $\{\mathbb{T}, \mathbb{X}, \mathbb{A}, \mathbb{S}\}$, where $\mathbb{T}$ denotes the time set, $\mathbb{X}$ is the state space, $\mathbb{A}$ is the set of initial states, and $\mathbb{S}$ denotes a family of motions. When $\mathbb{T} = \mathbb{R}^+ = [0, \infty)$, the tuple represents a continuous-time dynamical system, and when $\mathbb{T} = \mathbb{Z}^+ = \{0, 1, 2, \dots\}$, it implies a discrete-time dynamical system. For any *motion* $x(., x_0, t_0) \in \mathbb{S}$, we have $x(0, x_0, t_0) = x_0 \in \mathbb{A} \subset \mathbb{X}$, and $x(t, x_0, t_0) \in \mathbb{X}$ for all $t \in [t_0, t_1) \cap \mathbb{T}$, $t_1 > t_0$, where t_1 may be finite or infinite. The set of motions $\mathbb{S}$ is obtained by varying (t_0, x_0) over $(\mathbb{T} \times \mathbb{A})$.

A dynamical system is said to be *autonomous*, if every $x(., x_0, t_0) \in \mathbb{S}$ is defined on $\mathbb{T} \cap [t_0, \infty)$ and if for each $x(., x_0, t_0) \in \mathbb{S}$ and for each τ such that $t_0 + \tau \in \mathbb{T}$, there exists a motion $x(., x_0, t_0 + \tau) \in \mathbb{S}$ such that $x(t + \tau, x_0, t_0 + \tau) = x(t, x_0, t_0)$ for all t and τ satisfying $t + \tau \in \mathbb{T}$. A set $\mathbb{M} \subset \mathbb{A}$ is said to be *invariant* with respect to the set of motions $\mathbb{S}$ if $x_0 \in \mathbb{M}$ implies that $x(t, x_0, t_0) \in \mathbb{M}$ for all $t \geq t_0$, for all $t_0 \in \mathbb{T}$, and for all $x(., x_0, t_0) \in \mathbb{S}$. A point $p \in \mathbb{X}$ is called an *equilibrium* for the dynamical system $\{\mathbb{T}, \mathbb{X}, \mathbb{A}, \mathbb{S}\}$ if the singleton $\{p\}$ is an invariant set with respect to the motions $\mathbb{S}$.

The term *stability* (more specifically, Lyapunov stability) usually refers to the qualitative behaviour of motions relative to an invariant set (or an equilibrium), whereas the term *boundedness* (more specifically, Lagrange stability) refers to the (global) boundedness properties of the motions of a dynamical system [96].

Several research works in the systems and control field have been centred around the stability theorems of Lyapunov's methods and Lasalle's invariance principle, see, e.g. [9, 53, 62, 70, 80, 111, 130, 143]. Recently, there have been developments for the computerised and networked control systems that extend the traditional methods with non-monotonic Lyapunov functions, see, e.g. [96, 179]. The Lagrange stability had been introduced before the time the Lyapunov's methods were becoming well known, but has not been widely applied to the theoretical research as only provides a boundedness property or a uniformly bounded system. Furthermore, the Lyapunov stability defines the stability around a system equilibrium point—in other words, the stability of the equilibria.

An autonomous system is called asymptotically stable around its equilibrium point at the origin if it satisfies the following two conditions:

1. Given any $\varepsilon > 0$, $\exists \delta_1 > 0$ such that if $\|x(t_0)\| < \delta_1$, then $\|x(t)\| < \varepsilon$ $\forall t > t_0$.
2. $\exists \delta_2 > 0$ such that if $\|x(t_0)\| < \delta_2$, then $x(t) \to 0$ as $t \to \infty$.

The first condition requires that the state trajectory can be confined to an arbitrarily small ball centred at the equilibrium point, and of radius ε, when released from an arbitrary initial condition in a ball of sufficiently small radius δ_1. This is called stability in the sense of Lyapunov. It is possible to have stability in the sense of Lyapunov without having asymptotic stability.

The stability analysis for the quadratic constraint approach is provided in Appendix A, wherein the *GDC stability* and the *Input-to-power-and-State* stabilisation (IpSS) are defined formally. The differences between the GDC stability and the Lyapunov stability, as well as between the IpSS and the ISS, are also given in Appendix A.

Notation: The additional notations in this section are as follows. $\|u_i\|_\infty$ denotes the supremum norm, possibly $+\infty$, of u_i, which is typically an input or an output. When only the restriction of a signal to an interval I is relevant, $\|u_I\|_\infty$, or just $\|u_I\|$, denotes the supremum norm of that restriction, for instance $\|u_{[0,\kappa]}\|_\infty$, where $I = [0, \kappa]$, and similarly for scalar variables with the notations $|\xi|_\infty$ and $|\xi_{[0,\kappa]}|_\infty$. The vector element j of a vector field f is denoted as $f_{[j]}$. The matrix element $[j, l]$ of a block matrix B is denoted as $B_{[j,l]}$. The class $\mathscr{K}$, $\mathscr{L}$ and $\mathscr{K}\mathscr{L}$ functions are defined in the following:

- A function $\gamma : \mathbb{R}_0^+ \to \mathbb{R}_0^+$ is called of class $\mathscr{K}$ if it is continuous, strictly increasing and $\gamma(0) = 0$, and is called of class $\mathscr{K}_\infty$ if, in addition, it is unbounded.
- A function $\gamma : \mathbb{R}_0^+ \to \mathbb{R}_0^+$ is called of class $\mathscr{L}$ if it is monotonically decreasing and $\lim_{t \to \infty} \gamma(t) = 0$.
- A function $\alpha : \mathbb{R}_0^+ \times \mathbb{R}_0^+ \to \mathbb{R}_0^+$ is called of class $\mathscr{K}\mathscr{L}$ if it is of class $\mathscr{K}_\infty$ in the first and strictly decreasing to zero in the second argument.

1.3.1 System Model

Consider a discrete-time system $\mathscr{S}$ of the form:

$$\mathscr{S} : x(k + 1) = f(x(k)) + B(x(k))u(k) + d(k), \tag{1.7}$$

where $x \in \mathbb{R}^n$, $u \in \mathbb{U} \subset \mathbb{R}^m$ are the state and control vectors, respectively; $f(x)$ is the vector field, $f : \mathbb{R}^n \to \mathbb{R}^n$; $B(x)$ is the matrix field, $B : \mathbb{R}^n \to \mathbb{R}^{n \times m}$; the elements of f and B, denoted as $f_{[i]}(x)$ and $B_{[i,j]}(x)$, $f_{[i]} : \mathbb{R}^n \to \mathbb{R}$, $B_{[i,j]} : \mathbb{R}^n \to \mathbb{R}$, are not necessarily continuous in x; $d(k)$ represents the persistent unknown, but bounded, state disturbance $d \in \mathbb{V} \subset \mathbb{R}^n$.

Without loss of generality, we assume here that $f(0) = 0$; i.e., the zero is an equilibrium point of the uncontrolled nominal system $\mathscr{S} : x(k + 1) = f(x(k))$.

1.3.2 Supply Rates with Compound Vectors

In this section, the compound input and output vectors are introduced to the supply-rate function. Firstly, define the two compound vectors

$$u_\triangle^T := [u^T \ d^T] \text{ and } x_\triangle^T := [x^T \ f^T]. \tag{1.8}$$

Let $\xi_\triangle(x_\triangle(k), u_\triangle(k))$ be a real-valued continuous function in $x_\triangle$ and $u_\triangle$,

$$\xi_\triangle : \mathbb{R}^{2n} \times \mathbb{R}^{n+m} \to \mathbb{R}, \ \xi_\triangle < \pm\infty.$$

We assume that for any $x(0) \in \mathbb{X}$, $x_\triangle(0)$ and $\{u \in \mathbb{U}\}$ are such that the motion of $\xi_\triangle$, $\xi_{\triangle(k)} := \xi_\triangle(x_\triangle(k), u_\triangle(k))$, satisfies the following bounded property:

$$\exists \beta_\triangle \in \mathbb{R}^+ : \sum_{k=0}^{\kappa} \left|\xi_{\triangle(k)}\right| \leqslant \beta_\triangle \text{ for } \kappa > 0. \tag{1.9}$$

Similarly, define a real-valued supply-rate function without disturbances, $\xi(x_\triangle(k), u(k))$,

$$\xi : \mathbb{R}^{2n} \times \mathbb{R}^m \to \mathbb{R}, \ \xi < \pm\infty.$$

Only u, but not the compound $u_\triangle$, is included in this supply-rate function. The bounded property of $\xi_{(k)} = \xi(x_\triangle(k), u(k))$ is also applicable, i.e.

$$\exists \beta \in \mathbb{R}^+ : \sum_{k=0}^{\kappa} \left|\xi_{(k)}\right| \leqslant \beta \text{ for } \kappa > 0. \tag{1.10}$$

Remark 1.1 For a discrete-time linear-time-invariant system of the form

$$\mathscr{S}_\ell : x(k+1) = A\,x(k) + B\,u(k) + d(k), \tag{1.11}$$

the supply-rate function w.r.t input and state vectors $\xi_\ell(x(k), u(k))$ is often considered for the controlled system. The function $\xi_\ell : \mathbb{R}^n \times \mathbb{R}^m \to \mathbb{R}$ is a real-valued continuous function in u and x.

Remark 1.2 For a general discrete-time nonlinear system of the form

$$\mathscr{S}_\theta : x(k+1) = g(x(k), u(k)) + d(k), \tag{1.12}$$

the general supply-rate function w.r.t. input and state vectors, $\xi(k, x(k), u(k))$ the function $\xi : \mathbb{Z}^+ \times \mathbb{R}^n \times \mathbb{R}^m \to \mathbb{R}$, is considered for the controlled system $\mathscr{S}_\theta$ and not necessarily a continuous function in x. The supply-rate function with compound vectors for an input-affine system, $\xi_\triangle(x_\triangle(k), u_\triangle(k))$, can be expressed as $\xi_\theta(k, x(k), u(k))$. And similarly, $\xi(x_\triangle(k), u(k))$ can be expressed as $\xi_\vartheta(k, x(k), u(k))$.

The general dissipativity constraint (GDC) for nonlinear systems using the compound vectors is then defined below.

Definition 1.1 The controlled motion $(x_\triangle(k), u(k))$ of $\mathscr{S}$ (1.7) is said to satisfy the general dissipativity constraint (GDC), practically, if there exists a function $\alpha(.)$ of class $\mathscr{K}\mathscr{L}$ and a small positive scalar ε_k, such that

$$|\xi_{(k)}| \leqslant \alpha(|\xi_{(0)}|, k-1) + \varepsilon_k \quad \forall k \in \mathbb{Z}^+, \tag{1.13}$$

where $\xi_{(k)} = \xi(x_\triangle(k), u(k))$.

In implementations, ε_k can be set to zero upon a tuning threshold of $|\xi_{(k)}|$ for the nominal controlled systems.

Definition 1.2 The nominal controlled motion $(x_\triangle(k), u(k))$ of $\mathscr{S}$ (1.7) without disturbance $(d(k) = 0)$ is said to satisfy the general dissipativity constraint, ultimately, if there exists a function $\alpha(.)$ of class $\mathscr{K}\mathscr{L}$ and a small non-negative scalar ε_k, such that for all $k \in \mathbb{Z}^+$,

$$|\xi_{(k)}| \leqslant \alpha(|\xi_{(0)}|, k-1) + \varepsilon_k, \quad \text{for } \lim_{k\to\infty} \varepsilon_k = 0, \tag{1.14}$$

where $\xi_{(k)} = \xi(x_\triangle(k), u(k))$.

The simplest implementable form of a GDC is the monotonically decreasing bounded function of the form

$$0 \leqslant \xi_{(k)} \leqslant \gamma \times \xi_{(k-1)} < +\infty, \ \gamma \in \mathbb{R}^+, \ \gamma < 1, \tag{1.15}$$

or

$$0 \geqslant \xi_{(k)} \geqslant \gamma \times \xi_{(k-1)} > -\infty, \ \gamma \in \mathbb{R}^+, \ \gamma < 1, \tag{1.16}$$

where $\xi_{(k)} = \xi(x_\triangle(k), u(k))$. The other form of GDC that uses the absolute value of $\xi_{(k)}$ is stated below.

Definition 1.3 The controlled motion $(x_\triangle(k), u(k))$ of $\mathscr{S}$ (1.7) is said to satisfy the energy-dissipation constraint (EDC), practically, if there exists a positive time-dependent scalar $\delta_k \in \mathbb{R}^+$, $\delta_k < 1$, and a small positive scaler ε_k, such that for all $k \in \mathbb{Z}^+$,

$$|\xi_{(k)}| \leqslant \delta_k \times |\xi_{(k-1)}| + \varepsilon_k, \tag{1.17}$$

where $\xi_{(k)} = \xi(x_\triangle(k), u(k))$.

For clarity, the $\mathcal{KL}$-bounded property of a function is defined here: A function $\zeta : \mathbb{R}^q \to \mathbb{R}^q$ is said to be $\mathcal{KL}$-bounded, if there exists a class $\mathcal{KL}$ function $\alpha(.,.)$, such that for all $\zeta(k) \in \mathbb{R}^q$, $k \in \mathbb{R}_0^+$, we have the inequality of $\|\zeta(k)\| \leqslant \alpha(\|\zeta(0)\|, k - 1)$.

For the robust stabilisation of $\mathcal{S}$ in this GDC method, the supply rate $\xi_{(k)}$ is required to be $\mathcal{KL}$-bounded as in the above GDC (1.13), or its variants in (1.14), (1.15), (1.16), or (1.17), in combination with the dissipative condition of $\mathcal{S}$ (1.7) with respect to the supply rate $\xi_{\Delta(k)}$.

Remark 1.3 The above definitions are also applicable to the nonlinear system $\mathcal{S}_\theta$ (1.12).

Next, the dissipative system with a parameterised supply-rate and a parameterised storage function is defined below.

Definition 1.4 The system $\mathcal{S}$ (1.7) is said to be dissipative with respect to the parameterised supply rate $\xi_\Delta \left(x_\Delta(k), u_\Delta(k) \big|_{\Gamma_\Delta} \right)$, where $\Gamma_\Delta := \{Q_\Delta, S_\Delta, R_\Delta\}$, if there exists a real-valued non-negative parameterised storage function $V(x|_P)$ (e.g., of the quadratic form $V(x|_P) := x^T P x$, $P \succ 0$), such that the following dissipation inequality is satisfied for all $u_\Delta(k)$ and all $k > 0$, irrespectively of the initial state $x_\Delta(0)$:

$$\Delta_\tau V \left(x(k), x(k-1) \big|_{P(k), P(k-1)} \right) \leqslant \xi_\Delta \left(x_\Delta(k), u_\Delta(k) \big|_{\Gamma_\Delta} \right), \quad 0 < \tau < 1, \quad (1.18)$$

where $\Delta_\tau V \left(x(k), x(k-1) \big|_{P(k), P(k-1)} \right) := V \left(x(k) \big|_{P(k)} \right) - \tau V \left(x(k-1) \big|_{P(k-1)} \right)$.

The GDC stability can be defined with the dissipation inequality, as follows:

Definition 1.5 The controlled motion $(x(k), u(k))$ of the nominal system $\mathcal{S}$ (1.12) with the vanishing disturbance $d(k) = 0$ is said to be *GDC stable* around the zero equilibrium with respect to the supply rate $\xi_\theta (x(k), u(k))$ and the real-valued non-negative, Lipschitz continuous, and radially unbounded, storage function $V (x(k))$, if the following inequalities are satisfied for all $u(k) \in \mathbb{R}^m$ and all $k > 0$, irrespectively of the initial state $x(0)$:

$$V (x(k)) - \tau V (x(k-1)) \leqslant |\xi_\theta (x(k), u(k))|, \quad 0 < \tau < 1, \quad (1.19)$$

and there exist some control sequences $\{u_c(k) \in \mathbb{U} \subset \mathbb{R}^m\}$ such that $|\xi_\theta (x(k), u_c(k))|$ is $\mathcal{KL}$-bounded.

Remark 1.4 The GDC and GDC stability are defined for a single controlled (or closed-loop) system.

Alternatively, the GDC stability can be defined with two inequalities of the controlled system, as follows:

Definition 1.6 The controlled motion $(x(k), u(k))$ of the nominal system $\mathscr{S}$ (1.12) with the vanishing disturbance $d(k) = 0$ is said to be *GDC stable* around the zero equilibrium with respect to the supply rate $\xi_\theta (x(k), u(k))$ and the real-valued non-negative, Lipschitz continuous, and radially unbounded, storage function $V(x(k))$, if there exist some control sequences $\{u(k) \in \mathbb{U} \subset \mathbb{R}^m\}$ such that the following inequalities are satisfied for all $k > 0$, irrespectively of the initial state $x(0)$:

$$V(x(k)) - \tau V(x(k-1)) \leqslant |\xi_\theta (x(k), u(k))|, \; 0 < \tau < 1, \; \text{and} \quad (1.20)$$
$$|\xi_\theta (x(k), u(k))| \leqslant \alpha\big(|\xi_\theta (x(0), u(0))|, k-1\big), \quad (1.21)$$

for some $\mathscr{K}\mathscr{L}$ function $\alpha(., .)$.

The asymptotic stability and Input-to-State Stability (ISS) are now recalled here based on definitions in [65, 144, 147].

1.3.3 Stability

Consider the controlled system $\mathscr{S}$ (1.7). In a control design problem, either with a closed-form control law, or with an open-loop optimisation of model predictive control, the control can be represented by $u = j(k, x_\triangle(k))$. The closed-loop system model of $\mathscr{S}$ (1.7) can thus be expressed as:

$$\mathscr{S}_c : x(k+1) = f(x(k)) + B(x(k))j(k, x_\triangle(k)) + d(k)$$
$$:= z(k, x_\triangle(k)) + d(k). \quad (1.22)$$

1. *Lyapunov-based*: The system $\mathscr{S}$ (1.7) with $u = 0$ and $d = 0$, or system $\mathscr{S}_c$ (1.22) with $d = 0$, is locally asymptotically stable (LAS) around the zero equilibrium, if there is $\rho > 0$ and a function α of class $\mathscr{K}\mathscr{L}$ such that, given any initial state $x(0) = x_0$, $\|x_0\| < \rho$, a solution exists for each $k > 0$, satisfying

$$\|x(k, x_0)\| \leqslant \alpha(\|x_0\|, k). \quad (1.23)$$

2. *Input-to-State*: [65, 144] The system $\mathscr{S}$ (1.7) with $u = 0$, or system $\mathscr{S}_c$ (1.22), is said to be input-to-state stable, if there is a function γ of class $\mathscr{K}$ and a function α of class $\mathscr{K}\mathscr{L}$, such that for each bounded $\{d(k)\}$ and each initial state $x(0) = x_0$, a solution exists for each $k > 0$, satisfying

$$\|x(k, x_0, d)\| \leqslant \alpha(\|x_0\|, k) + \gamma(\|d\|_\infty). \quad (1.24)$$

The inequality (1.24) in the definition of ISS can also be written as

$$\|x(k, x_0, d)\| \leqslant \max\big\{\alpha(\|x_0\|, k), \; \gamma(\|d\|_\infty)\big\}. \quad (1.25)$$

ISS gain property: When a system is input-to-state stable, it has an asymptotic gain property of the form

$$\overline{\lim_{k \to \infty}} \left(\|x(k, x_0, d)\| \right) \leqslant \gamma(\|d\|_\infty).$$

Also, a system is ISS if and only if there exists an ISS Lyapunov function, and a local ISS $\Longrightarrow$ LAS [147], but not reversely. Further characteristics of ISS are given in [65, 146].

3. *Bounded Input Bounded Output*: An LTI system is Bounded Input Bounded Output (BIBO) stable if every bounded input $u(t)$ produces a bounded output $y(t)$. The SISO linear system with impulse response $h(t)$ is BIBO stable if and only if $h(t)$ is absolutely integrable, i.e. $\exists \mu < \infty$ such that

$$\int_0^\infty |h(\tau)| d\tau < \mu.$$

Remark 1.5 For clarity, it is worth providing the definitions for Lyapunov stability, asymptotic attractivity and asymptotic stability herein.

Consider a continuous-time dynamical system $\mathscr{S} : \dot{x}(t) = f(x(t))$, $t \in \mathbb{R}$, $x \in \mathbb{R}^n$, where the function $f : \mathbb{R}^n \to \mathbb{R}^n$ is uniformly continuous, $f(0) = 0$.

1. *Lyapunov stability*: The origin of $\mathscr{S}$ is said to be Lyapunov stable if $\forall \varepsilon \in \mathbb{R}_0^+$ and $\forall t_0 \in \mathbb{R}$ there exists $\delta = \delta(\varepsilon, t_0) \in \mathbb{R}_0^+$, such that a solution $x(t, t_0, x_0) \in \mathscr{B}(\varepsilon)$ exists $\forall t > t_0$, if $x_0 \in \mathscr{B}(\delta)$.
2. *Asymptotic attractivity*: The origin of $\mathscr{S}$ is said to be asymptotically attractive if $\forall t_0 \in \mathbb{R}$ there exists a set $\mathscr{X}(t_0) \subseteq \mathbb{R}^n : 0 \in \mathscr{X}(t_0)$, such that a solution $x(t, t_0, x_0)$ exists for $x_0 \in \mathscr{X}(t_0)$, and $\lim_{t \to +\infty} \|x(t, t_0, x_0)\| = 0$.
3. *Asymptotic stability in Lyapunov sense*: The origin of $\mathscr{S}$ is said to be asymptotically stable in Lyapunov sense if it is Lyapunov stable and asymptotically attractive.

Some results for linear systems:

A continuous-time linear system is asymptotically stable if and only if real parts of all poles (or eigenvalues of the system matrix) are negative. A linear system is marginally stable if and only if it has at least one simple pole (not repeated) with real part zero, and all other poles have negative real parts. A discrete-time linear system is asymptotically stable if and only if all eigenvalues of the system matrix are strictly inside the unit circle.

If a linear system is asymptotically stable, then it is BIBO stable. If a linear system is BIBO stable and the state-space representation is minimal, i.e. both controllable and observable, then the system is asymptotically stable. A marginally stable system with minimal realisation is not BIBO stable.

For a constructive control design procedure, it is necessary to make some assumptions on the positive invariance of $\mathbb{X}$ ($x(k) \in \mathbb{X}$) in order for the problem to be feasible.

Definition 1.7 A set $\mathbb{X} \subset \mathbb{R}^n$ is called robustly constrained control invariant with respect to $\mathbb{U}$ and $\mathbb{V}$ of the system $\mathscr{S}$ if for each $x_k \in \mathbb{X}$, $\exists u_k \in \mathbb{U}$, such that $x_{k+1} \in \mathbb{X}$ for all $d_k \in \mathbb{V}$.

The readers may refer to [12] for further applications of invariant sets in the control theory and practice. The reader may also refer to a recent review paper on robust control [110].

1.3.4 Passivity and Small-Gain Theorems

The passivity theorem, small-gain theorem and the method for control synthesis and design with $\mathscr{L}_2$ gain [166] and dissipative and passive systems in the literature are briefly recapped here. This line of research has been mathematical oriented. The methods have been developed and presented by mathematicians and academics from the mathematical schools.

The passivity and dissipativity are the input–output properties of a system, or operator $H_i : u_i \mapsto H_i(u_i) = y_i$. The system is assumed to be well posed as an input–output system; i.e., it is assumed that $H_i : \mathscr{L}_{2e} \to \mathscr{L}_{2e}$. The definitions for $\mathscr{L}_q$ signal space and the extended $\mathscr{L}_q$ signal space with the truncation, $\mathscr{L}_{qe}$, are given in [166].

Theorem 1.1 ([17] Theorem 5.80, pp. 310–311)—*Small-gain theorem*
Consider a negative feedback interconnection of two systems, H_1 and H_2, with the corresponding perturbations r_1 and r_2, as in Fig. 1.13.

Fig. 1.13 Feedback interconnection type 1 [17]

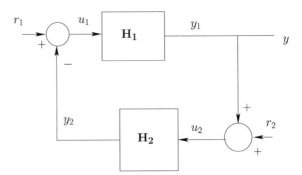

Suppose that the two systems H_1 and H_2 are both causal and with finite power gain γ_1 and γ_2 [166], and power bias λ_1 and λ_2, respectively, assuming $u_i(t) \in \mathscr{L}_{2e}$, $y_i(t) \in \mathscr{L}_{2e}$, $i = 1, 2$.

If $\gamma_1 \gamma_2 < 1$ then for all inputs $||r_1||_{fp} < +\infty$ and $||r_2||_{fp} < +\infty$, the closed-loop interconnection is stable in the sense that $||u_1||_{fp} < +\infty$, $||u_2||_{fp} < +\infty$, $||y_1||_{fp} < +\infty$, $||y_2||_{fp} < +\infty$, i.e. has $\mathscr{L}_2$ gain,

$$where \ ||v||_{fp} := \sqrt{\lim_{t \to +\infty} \sup \left\{ \frac{1}{t} \int_0^t v^T(\tau) v(\tau) d\tau \right\}}.$$

There are different versions of small-gain theorems. The small-gain theorems with input-to-state stability have also been stated elsewhere, see, e.g., [62, 85]. The storage functions have not been considered for systems having $\mathscr{L}_2$ gains in this line of research. The definitions for passive and dissipative systems are in the following:

Definition 1.8 A system H with input $u(.)$ and output $y(.)$ where $u(t), y(t) \in \mathbb{R}^m$ is passive if there is a constant β such that

$$\int_0^t y^T(\tau) u(\tau) d\tau \geq \beta \tag{1.26}$$

for all functions $u(.)$ and all $t \geq 0$. If, in addition, there are constants $\delta \geq 0$ and $\varepsilon \geq 0$ such that

$$\int_0^t y^T(\tau) u(\tau) d\tau \geq \beta + \delta \int_0^t u^T(\tau) u(\tau) d\tau + \varepsilon \int_0^t y^T(\tau) y(\tau) d\tau \tag{1.27}$$

for all functions $u(.)$, and all $t \geq 0$, then the system is input strictly passive (ISP) if $\delta > 0$, output strictly passive (OSP) if $\varepsilon > 0$, and very strictly passive (VSP) if $\delta > 0$ and $\varepsilon > 0$.

Now, consider the nonlinear state-space system, given by

$$\Sigma : \begin{cases} \dot{x} = f(x, u), \\ y = h(x, u), \end{cases} \tag{1.28}$$

where $x \in \mathbb{X} \subseteq \mathbb{R}^n$, $u \in \mathbb{U} \subseteq \mathbb{R}^m$, and $y \in \mathbb{Y} \subseteq \mathbb{R}^p$. Associated with (1.28) is a real-valued supply-rate function of $w(u, y) : \mathbb{U} \times \mathbb{Y} \to \mathbb{R}$.

Definition 1.9 (*Willems J., 1972*, [172]) The state-space system (1.28) is said to be dissipative with respect to the supply rate $w(u, y)$, if there exists a real-valued function $V : \mathbb{X} \to \mathbb{R}_0^+$, called the storage function, such that $\forall x_0 \in \mathbb{X}$, $\forall t_1 > t_0$, and all admissible inputs u the following dissipation inequality holds

$$V(x(t_1)) \leq V(x(t_0)) + \int_{t_0}^{t_1} w(u(t), y(t)) dt, \tag{1.29}$$

where $x(t_0) = x_0$, and $x(t_1)$ is the state of the supply rate at time t_1 resulting from initial state x_0 and the input function $u(t)$.

In the following theorem, consider two dynamical systems of the form

$$\dot{x}_i(t) = f_i(x_i(t), u_i(t)), \ \ y_i(t) = h_i(x_i(t)), \ \ i = 1, 2,$$

which have uniform finite power gains $\gamma_i \geqslant 0$, if they are quasi-dissipative with supply rate $w_i(u_i, y_i) = \gamma_i^2 u_i^T u_i - y_i^T y_i$, i.e. $\int_0^t w_i(u_i(\tau), y_i(\tau))d\tau \geqslant \beta, \ \beta \leqslant 0$.

Theorem 1.2 ([17] Proposition 5.8, pp. 258–262)—*Passivity theorem*
Suppose that the systems H_1 and H_2 are quasi-dissipative with respect to supply rates $w_1(u_1, y_1)$ and $w_2(u_2, y_2)$, respectively. Suppose there exists $\rho > 0$ such that the matrix

$$\begin{bmatrix} Q_1 + \rho R2 & -S1 + \rho S_2^T \\ -S_1^T + \rho S_2 & R_1 + \rho Q_2 \end{bmatrix}$$

is negative definite. Then the feedback system formed by the interconnection of two systems, H_1 and H_2, with the corresponding perturbations r_1 and r_2, as in Fig. 1.13, has uniform finite power gain.

In a passivity theorem, the problem is always assumed well posed; i.e., all the signals belong to $\mathscr{L}_{2e}$.

The following definitions are from [139]. They are for one-port nonlinear systems H in Fig. 1.14 of the form:

$$(H): \quad \begin{aligned} \dot{x} &= f(x, u), \quad x \in \mathbb{R}^n, \\ y &= h(x, u), \quad u, y \in \mathbb{R}^m, \end{aligned} \tag{1.30}$$

where the state $x(t)$ is uniquely determined by its initial value $x(0)$ and the input $u(t) \in \mathbb{U}$; $f(0, 0) = 0$ and $h(0, 0) = 0$.

Theorem 1.3 ([139] Theorem 2.30, pp. 32–33)—*Feedback interconnection of passive systems*
Suppose that H_1 and H_2 are passive. Then the system obtained by the feedback interconnection, as in Fig. 1.15, is passive.

The following *zero-state detectable* property [139] for nonlinear systems is important for dissipative systems with inputs and outputs.

Fig. 1.14 One port
system—a block diagram
[139]

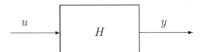

Fig. 1.15 Feedback
interconnection type 2 [139]

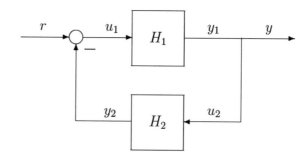

Since the storage function is only positive semi-definite, the zero state $x = 0$ can be unstable when there is an unobservable part. In linear systems, the stability is achieved with a '*detectability*' assumption, which requires that the unobservable part of the system be asymptotically stable. The definition below is for nonlinear systems.

Definition 1.10 ([139] *Definition 2.27, pp. 48–49)—Zero-state detectability and zero-state observability*
Consider the system H with zero input, that is $x = f(x, 0)$, $y = h(x, 0)$, and let $\mathbb{W} \subset \mathbb{R}^n$ be its largest positively invariant set contained in $\{x \in \mathbb{R}^n \mid y = h(x, 0) = 0\}$. We say that H is *zero-state detectable* (ZSD) if $x = 0$ is asymptotically stable conditionally to $\mathbb{W}$. If $\mathbb{W} = \{0\}$, we say that H is *zero-state observable* (ZSO).

Whenever the ZSD property is used to establish a global result, it is assumed that $x = 0$ is globally asymptotically stable (GAS), conditionally to $\mathbb{W}$.

Theorem 1.4 ([139] *Theorem 2.28, pp. 49–50)—Passivity and stability*
Let the system H be passive with a C^1 storage function S and $h(x, u)$ be C^1 in u for all x. Then, the following properties hold:

1. If S is positive definite, then the equilibrium $x = 0$ of H with $u = 0$ is stable.
2. If H is ZSD, then the equilibrium $x = 0$ of H with $u = 0$ is stable.
3. When there is no throughput, $y = h(x)$, then the feedback $u = -y$ achieves asymptotic stability of $x = 0$ if and only if H is ZSD.

Theorem 1.5 ([139] *Theorem 2.10, pp. 50–51)—Feedback interconnection of passive systems*
Assume that the systems H_1 and H_2 are dissipative with the supply rates

$$w_i(u_i, y_i) = u_i^T y_i - \rho_i(y_i)y_i - \nu_i^T(u_i)u_i, \quad i = 1, 2,$$

where $\rho_i(.)$ and $\nu_i(.)$ are nonlinear functions. Furthermore, assume that they are zero-state detectable and that their respective storage functions $S_1(x_1)$ and $S_2(x_2)$ are C^1. Then, the equilibrium $(x_1, x_2) = (0, 0)$ of the feedback interconnection with the reference $r \equiv 0$, as shown in Fig. 1.15, is asymptotically stable, if
$\nu_1^T(v) + \rho_2^T(v) > 0$ and $\nu_2^T(v) + \rho_1(v) > 0$ for all $v \in \mathbb{R}^m \setminus \{0\}$.

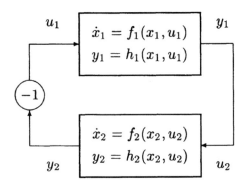

Similar theorems have been stated in [53, 62] for the asymptotic stability.

Theorem 1.6 ([62] Theorem 10.8.1, pp. 55–57)—*Feedback interconnection of dissipative systems*
    Consider the feedback interconnection of two dissipative systems Σ_1 and Σ_2, as in Fig. 1.16. Let two supply rates

$$w_i(u_i, y_i) = u_i^T R_i u_i + 2 y_i^T S_i u_i + y_i^T Q_i y_i, \ i = 1, 2$$

be given and suppose that, for some $\delta > 0$, the symmetric matrix

$$M = \begin{bmatrix} Q_1 + \delta R2 & -S1 + \delta S_2^T \\ -S_1^T + \delta S_2 & R_1 + \delta Q_2 \end{bmatrix}$$

is either:

1. *negative semi-definite,*
2. *negative semi-definite, $S_1 - \delta S_2^T$ and $R_l + \delta Q_2$ are nonsingular,*
3. *negative definite;*

    Then, if both Σ_1 and Σ_2 are strictly dissipative and (1) holds, the negative feedback interconnection is globally asymptotically stable. If both Σ_1 and Σ_2 are dissipative, (2) holds, and $\Sigma_2 \circ \Sigma_1$ is zero-state detectable, the negative feedback interconnection is globally asymptotically stable. If both Σ_1 and Σ_2 are dissipative, (3) holds, and both Σ_1 and Σ_2 are zero-state detectable, the negative feedback interconnection is globally asymptotically stable.

The next theorem is for discontinuous systems.

Theorem 1.7 ([53] Theorem 5.2, pp. 104–105)—*Feedback interconnection of dissipative discontinuous systems*
    Consider the feedback interconnection of two 'strongly' dissipative systems, Σ_1 and Σ_2, w.r.t the two supply rates $w_i(u_i, y_i) = u_i^T R_i u_i + 2 y_i^T S_i u_i + y_i^T Q_i y_i, \ i = 1, 2$, as in Fig. 1.16, in which $f_1(x_1, u_1) = g_1(x_1) + G_1(x_1)u_1, \ f_2(x_2, u_2) = g_2(x_2) + $

$G_2(x_2, u_2)u_2$, $h_1(x_1, u_1) = l_1(x_1) + L_1(x_1)u_1$, $h_2(x_2, u_2) = l_2(x_2) + L_2(x_2, u_2)u_2$, *satisfying the following assumptions: $g_i(.)$, $G_i(., .)$, $l_i(.)$, $L_i(., .)$ are Lebesgue measurable and locally essentially bounded, Σ_2 has at least an equilibrium point, and the feedback interconnection is well posed, that is $det[I_m + L_2(x_2, y_1)L_1(x_1)] \neq 0$ for all, y_1, x_1, x_2.*

Suppose that, for some $\delta > 0$,

$$M = \begin{bmatrix} Q_1 + \delta R2 & -S1 + \delta S_2^T \\ -S_1^T + \delta S_2 & R_1 + \delta Q_2 \end{bmatrix} \leqslant 0,$$

and both storage functions are locally Lipschitz continuous, regular and radially unbounded. Then, the following statements hold:

1. *The negative feedback interconnection of Σ_1 and Σ_2 is Lyapunov stable.*
2. *If $M < 0$, then the negative feedback interconnection of Σ_1 and Σ_2 is globally asymptotically stable.*

Discussion: The storage functions have always been employed in the passivity theorems stating the feedback interconnection stability conditions. In the work of Willems J., see, e.g., [173], a general dissipative system is defined as having the following dissipativity inequality, as a system property, fulfilled by every $u(t)$ and $y(t)$ satisfying the system state-space equations:

$$-\int_0^T w(u(t), y(t))dt \geqslant \beta \text{ for } T \geqslant 0.$$

The small-gain theorems were originally associated with the dissipativity inequality without a storage function, while the passivity theorems are often associated with two dissipative systems having storage functions such that the following dissipation inequalities hold along all possible trajectories:

$$\int_0^t \xi_i(u_i(\tau), y_i(\tau))d\tau \geqslant V_i(x_i(t)) - V_i(x_i(0)), \text{ for all } x_i(0), \ i = 1, 2, \ t \geqslant 0.$$

The small-gain theorems with some storage functions have also been developed, but some recent works have found that the small-gain theorems may become conservative in the decentralised control problems, e.g. [31].

The presenting GDC method in this book, however, employed both of the above inequalities: the one with a storage function is for the plant, and the one without any storage functions is for the controller in a closed loop with that plant as a single system—the closed-loop system perspective. Therefore, the interconnection stability condition will not be presented in this book.

Quadratic Difference Forms: The Quadratic Differential Form (QDF) [114] has been a part of the behavioural framework for dynamical systems. A stability condition for the discrete-time version of the QDF is provided below. Define

$$\hat{u}(t) := [u^T(t) \; u^T(t+1) \; \ldots \; u^T(t+n)]^T,$$

$$\hat{y}(t) := [y^T(t) \; y^T(t+1) \; \ldots \; y^T(t+m)]^T,$$

for some finite n and m, and $w(t) \triangleq [\hat{y}^T(t), \; \hat{u}^T(t)]^T$, a discrete-time QDF can be defined in a compact form as

$$\Psi_\phi(y, u) := \sum_{j=0}^{k} \sum_{i=0}^{k} w^T(t+j) \, \phi_{ji} \, w(t+j),$$

where k is called the degree of supply rate and can be the current time step.

Theorem 1.8 ([72] - Kojima and Takaba, 2005) *The zero of a discrete-time linear-time-invariant (LTI) system is asymptotically stable if there exists a symmetric two-variable polynomial matrix*

$$\Psi_\phi(\zeta, \eta) \geqslant 0 \text{ and } \triangle \Psi_\phi > 0.$$

for all possible inputs and outputs satisfying the system equations.

1.3.5 Nucleus Contributions

In this book, we introduce a novel method for designing the distributed and decentralised controllers of interconnected and cyber-physical systems in the discrete-time domain on the grounds of *'general dissipativity constraint'* (GDC), closed-loop dissipative system, relaxed non-monotonic Lyapunov function and an extension of input-to-state stability. The theorems in this chapter are not applicable and will not be used in the presented *'GDC method'* in this book. The non-centralised *model predictive control* (MPC) of interconnected systems in various application domains and the *'dependable control system'* have been determined (but are not limited to) as the potential applications of the GDC method.

The interconnection of $\mathscr{L}_2$–gain and dissipative open-loop systems with the small-gain theorems and passivity theorems, as recapped above, are well established in the literature. In those theorems, the asymptotic stability of the zero equilibrium has been obtained essentially from having $\dot{V}_t \leqslant 0$ with an appropriately radially unbounded Lyapunov function $V(x)$. The third method is associated with the 'quadratic differential forms' for the closed-loop systems within the behavioural framework.

The presented method with the GDC can be considered as *the fourth one*, but in the discrete-time domain only. In the GDC method, the asymptotic attractivity is obtained from having $\triangle V_k \geqslant 0$ and decreasing, but not necessarily monotonically (in contrast to the discrete-time version of $\dot{V}_t \leqslant 0$). We have developed the sufficient conditions for asymptotic attractivity and *'stability in the GDC sense'* and also stated the theorems for continuous and discontinuous discrete-time systems. Both the storage function and the supply-rate function are generally piece-wise continuous in time with the GDC method. The storage function is only a relaxed non-monotonic Lyapunov function.

When applying to the model predictive control problem, the asymptotic stability is usually obtained whenever the recursive feasibility of the MPC optimisation is assured, or guaranteed, at every time step. For the decentralised MPC problem, the newly introduced *'GDC attractability'* is assumed for obtaining the stability in the GDC sense in a decentralised control architecture, which can be translated to the conditions on the guaranteed recursive feasibility for every local MPC optimisation. The GDC attractability for a decentralised control scheme is usually stronger than the combined controllability and observability of the global system as a single stand-alone system.

The *enforced attractivity constraints* for a partially and fully decentralised MPC scheme and the *stabilising agent* segregated from the companion control algorithm are the two main implementations of the GDC method in this book. The GDC method is applicable to nonlinear systems in general, but the numerical examples are essentially of MPC and quadratic supply functions for linear systems.

Chapter 2
Quadratic Constraint for Decentralised Model Predictive Control

The asymptotically positive realness constraint (APRC) and quadratic dissipativity constraint (QDC) are introduced in this chapter as an effective tool for designing decentralised control systems, especially the decentralised model predictive control, in the discrete-time domain. We derive the convergence conditions for interconnected systems on the grounds of global system dissipation, subsystem dissipation, coupling structure, and dissipation-based constraints (APRC or QDC in the case of quadratic constraints) of all controlled subsystems. These conditions are suitable for the decentralised and distributed control of interconnected systems that prohibit artificial constraints on the unmeasurable coupling vectors. A convex quadratic constraint on the current-time control vector is subsequently developed from the dissipation-based constraint and applied to the model predictive control (MPC) as an enforced attractivity constraint. The attractivity constraints for controlled subsystems can be fully decoupled in this approach. Only linear-time-invariant (LTI) interconnected systems are under the scope of this chapter.

2.1 Control and System Models

The interaction-oriented model [86] is employed in this section. Consider an interconnected system Σ consisting of h subsystems, each denoted as $\mathscr{S}_i$, $i = 1, 2, \ldots, h$, and has a discrete-time state-space model of the following form:

$$\mathscr{S}_i : \begin{cases} x_i(k+1) = A_i x_i(k) + B_i u_i(k) + E_i v_i(k), \\ y_i(k) = C_i x_i(k), \\ w_i(k) = F_i x_i(k), \end{cases} \tag{2.1}$$

© Springer Nature Singapore Pte Ltd. 2018
A. Tri Tran C. and Q. Ha, *A Quadratic Constraint Approach to Model Predictive Control of Interconnected Systems*, Studies in Systems, Decision and Control 148, https://doi.org/10.1007/978-981-10-8409-6_2

where $u_i \in \mathbb{U}_i \subset \mathbb{R}^{m_i}$, $y_i \in \mathbb{R}^{q_i}$, $x_i \in \mathbb{R}^{n_i}$, $v_i \in \mathbb{R}^{m_{v_i}}$, $w_i \in \mathbb{R}^{q_{w_i}}$; x_i, u_i, y_i, v_i, and w_i are, respectively, state, control, output, interactive (or coupling) input and output vectors. In this chapter, the following ball constraint is considered

$$\mathbb{U}_i = \{u_i : \|u_i - u_i^{ss}\|^2 \leqslant \eta_i\}, \tag{2.2}$$

where u_i^{ss} is the steady state of $u_i(k)$.

When $v_i(k) = w_j(k)$, $j \neq i$, we say that the subsystems $\mathscr{S}_i$ and $\mathscr{S}_j$ are linearly connected to each other. The subsystem connections are encapsulated by the global interconnection matrix H, also called global coupling matrix. The elements of H are either 1 or 0 only, with 1 representing a connection.

The larger-scale global system Σ is represented by a state-space model of the block-diagonal system $\mathscr{S}$ formed by h subsystems $\mathscr{S}_i$, $i = 1, 2, \ldots, h$, and the global interconnection matrix H, as follows:

$$\Sigma : \begin{cases} x(k+1) = A\,x(k) + B\,u(k) + E\,v(k), \\ \quad y(k) = C\,x(k), \\ \quad w(k) = F\,x(k), \\ \quad v(k) = H\,w(k), \end{cases} \tag{2.3}$$

where $A = \mathrm{diag}[A_i]_1^h$, $x = [x_1^T \ldots x_h^T]^T$, and similarly for other global block-diagonal matrices B, C, E, F and the stacking vectors u, y, v, w of Σ. From a centralised point of view, (2.3) can be rewritten as

$$\Sigma : \begin{cases} x(k+1) = (A + EHF)\,x(k) + B\,u(k) = A_\Sigma\,x(k) + B\,u(k), \\ \quad y(k) = C\,x(k). \end{cases} \tag{2.4}$$

In a decentralised control scheme, the models of stand-alone subsystems are required by the local control algorithms. When all the interactive inputs and outputs vanish (i.e. $v_i = 0$ and $w_i = 0$), a stand-alone subsystem $\mathscr{S}_i$ has the state-space model of the form:

$$\mathscr{S}_i|_{\text{stand alone}} : \begin{cases} x_i(k+1) = A_i x_i(k) + B_i u_i(k), \\ \quad y_i(k) = C_i x_i(k). \end{cases} \tag{2.5}$$

The following properties are assumed in this work:

1. (A, C) and (A_i, C_i) are observable while (A, B) and (A_i, B_i) are controllable.
2. The updating time instants are synchronised between all subsystems and their local controllers.

Problem 1 statement:

1. To implement the conventional model predictive control (MPC) algorithm [88] using the stand-alone model (2.5) in the prediction for each single subsystem $\mathscr{S}_i$

(2.1), i.e. the interactive variables v_i and w_i vanish in the objective function of each local MPC. The global system may be open-loop unstable. There are not any communication links and exchanged-information data between subsystem controllers.

2. To design an enforced *attractivity constraint* for each local MPC to guarantee that the plant-wide global system is asymptotically attractive. In other words, we are concerned with the design of h decoupled attractivity constraints for the local MPCs of h associated subsystems $\mathscr{S}_i$, such that the global system Σ (2.4) is asymptotically attractive. Only local control constraint (2.2) is considered in the MPC optimisation. The term 'attractivity constraint' comes from the term 'stability constraint' for MPC in the literature [93], to imply that only the asymptotic attractivity is guaranteed by the dissipation-based constraint.

3. To determine the control performance via the local MPC cost functions, which are chosen by the application on case by case basis. In other words, the weighting coefficients of the MPC cost function are user-chosen parameters which are dependent upon the other operational performances.

2.2 Asymptotic Attractivity Condition

This section presents the asymptotically attractive conditions for the controlled interconnected system Σ with a decentralised MPC. The dissipation-based constraint is converted into an enforced attractivity constraint for the MPC, which is an additional constraint on the current-time control vector. The interconnection stability condition as in the dissipative system theory that is established by the connection of a dissipative system and a strictly dissipative system is not employed in this approach. The asymptotic attractivity of the global system associated with a decentralised control scheme is guaranteed from the satisfaction of dissipative criteria and the dissipation-based constraints of all controlled subsystems, together with a global condition on the coefficient matrices related to the interactive (or coupling) variables.

2.2.1 Quadratic Constraint

Consider a quadratic supply-rate function for each controlled subsystem $\mathscr{S}_i$ in the following:

$$\xi_i\big(u_i(k), y_i(k)\big) := \begin{bmatrix} u_i(k)^T & y_i(k)^T \end{bmatrix} \begin{bmatrix} Q_i & S_i \\ * & R_i \end{bmatrix} \begin{bmatrix} u_i(k) \\ y_i(k) \end{bmatrix}, \qquad (2.6)$$

where Q_i, R_i, S_i are coefficient matrices with symmetric Q_i and R_i, $Q_i \in \mathbb{R}^{q_i \times q_i}$, $R_i \in \mathbb{R}^{m_i \times m_i}$, $S_i \in \mathbb{R}^{q_i \times m_i}$.

Definition 2.1 The controlled motion $(y_i(k), u_i(k))$ of $\mathscr{S}_i$ (2.1) is said to satisfy the *positive realness constraint* (PRC) with respect to $\xi_i(u_i(k), y_i(k))$, if

$$\exists k_0 > 0: \; \xi_i(u_i(k), y_i(k)) \geqslant 0 \; \forall k > k_0. \tag{2.7}$$

Definition 2.2 The controlled motion $(y_i(k), u_i(k))$ of $\mathscr{S}_i$ (2.1) is said to be *negatively asymptotically attractive* with respect to $\xi_i(u_i(k), y_i(k))$, if $\xi_i(u_i(k), y_i(k))$ is non-positive and finite for all $k > 0$, and

$$\xi_i \to 0 \text{ as } k \to +\infty. \tag{2.8}$$

The positive realness constraint can be sufficiently obtained by an incremental condition in the following lemma.

Lemma 2.1 *The controlled motion $(y_i(k), u_i(k))$ of $\mathscr{S}_i$ (2.1) satisfies the positive realness constraint with respect to $\xi(u_i(k), y_i(k))$, $\xi_i(u_i(0), y_i(0)) < 0$, if there exists a real number γ_i, $0 < \gamma_i < 1$, and a positive real-valued function $\mu_i(k) : \mathbb{Z}^+ \to \mathbb{R}^+$, such that the following asymptotically positive realness constraint (APRC) is satisfied:*

$$\exists k_s > 0: \; 0 \geqslant \xi_i(u_i(k), y_i(k)) \geqslant \gamma_i \times \xi_i(u_i(k-1), y_i(k-1)) + \mu_i(k) \; \forall k \in (0, k_s], \tag{2.9}$$

in which $\mu_i(k_s) + \gamma_i \times \xi_i(u_i(k_s - 1), y_i(k_s - 1)) > 0$.

For conciseness, $\xi_i(u_i(k), y_i(k))$ is denoted as ξ_i. And the time index k is omitted, while $k + 1$ and $k - 1$ are denoted by the superscripts $^+$ and $^-$, respectively, where appropriate. The negatively asymptotic attractiveness can be sufficiently obtained by an incremental condition in the following lemma.

Lemma 2.2 *The controlled motion $(y_i(k), u_i(k))$ of $\mathscr{S}_i$ (2.1) is negatively asymptotically attractive with respect to $\xi(u_i(k), y_i(k))$, if there exists a real number γ_i, $0 < \gamma_i < 1$, such that the following quadratic dissipativity constraint (QDC) holds:*

$$0 \geqslant \xi_i \geqslant \gamma_i \times \xi_i^- \; \forall k > 0. \tag{2.10}$$

Proof From (2.10), we obtain by induction

$$0 \geqslant \xi_i \geqslant -\gamma_i^k \times o_i,$$

where $o_i = \left| \xi_i(u_i(0), y_i(0)) \right| > 0$. Therefore, $\xi_i \to 0$ as $k \to +\infty$, due to $\gamma_i < 1$ ∎

Remark 2.1 The PRC is often a non-convex constraint in the variable u_i, thus will not be employed in a design procedure in this chapter. As an alternative, we can combine the PRC with a dissipative criterion such that the asymptotic stability will be obtained with a Lyapunov function (starting from the time step k_s), then the convergence of $\xi_i \to 0$ will be obtained accordingly.

Remark 2.2 When the QDC is combined with a dissipative criterion, the convergence of both the supply rate and the state can be obtained simultaneously.

The dissipation-based constraint (APRC or the QDC) can be converted into an inequality constraint in $u_i(k)$ when $y_i(k)$ and ξ_i^- are known a priori, which can then be imposed onto the optimisation problem of MPC as an extra constraint for assuring the controlled system attractivity.

In what follows, the QDC inequality (2.10) will be used to derive the attractivity constraint.

2.2.2 Attractivity Constraint and Its Qualification

Rewrite the second inequality of (2.10) as

$$\boldsymbol{u}_i(k)^T Q_i \boldsymbol{u}_i + 2 y_i(k)^T S_i^T \boldsymbol{u}_i(k) \geqslant \gamma_i \times \xi_i^- - y_i(k)^T R_i y_i(k), \qquad (2.11)$$

in which $y_i(k)$ and ξ_i^- are known.

This is an inequality constraint in the variable $\boldsymbol{u}_i(k)$ and convex when $Q_i \prec 0$, which can be applied to the MPC optimisation as an enforced attractivity constraint. An example of the QDC-based attractivity constraint is given in Fig. 2.1, in which the ellipsoids governed by (2.11) change along the trajectory. This is from the automatic generation control (AGC) application of a power system in a numerical example.

For systems having the control constraint $u_i \in \mathbb{U}_i$, this enforced attractivity constraint must meet an extra qualification condition to ensure that there exists $\boldsymbol{u}_i(k) \in \mathbb{U}_i$ that fulfils (2.11).

Fig. 2.1 Control trajectory and QDC-based attractivity constraints

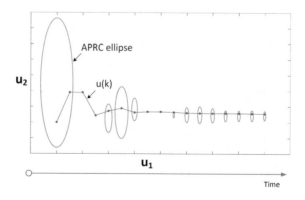

(a) *Control qualification condition 1*

For $Q_i \prec 0$, $u_i^{ss} = 0$, and $0 \in \mathbb{U}_i$, the ellipsoid region governed by (2.11) intersects $\mathbb{U}_i$ if the right-hand side (RHS) of (2.11) is negative since (2.11) will hold true with $u_i(k) = 0$, i.e.

$$\gamma_i \times \xi_i^- - y_i^T(k) R_i y_i(k) < 0. \tag{2.12}$$

(b) *Control qualification condition 2*

By rewriting

$$\xi_i = (u_i + (Q_i)^{-1} S_i y_i)^T Q_i (u_i + (Q_i)^{-1} S_i y_i) + y_i^T (R_i - S_i^T (Q_i)^{-1} S_i) y_i,$$

Equation (2.11) can be expressed by the following ellipsoid inequality:

$$(u_i + (Q_i)^{-1} S_i y_i)^T Q_i (u_i + (Q_i)^{-1} S_i y_i) \geqslant \gamma_i \times \xi_i^- - y_i^T (R_i - S_i^T (Q_i)^{-1} S_i) y_i. \tag{2.13}$$

The feasibility of (2.27) is thus guaranteed, i.e. the QDC is *control-constrained qualified*, if the RHS of (2.13) is negative for $Q \prec 0$, and the centre point of the ellipsoid governed by (2.13) belongs to $\mathbb{U}_i$.

Lemma 2.3 *For $\tilde{Q}_i = (Q_i)^{-1} \prec 0$, suppose that*

$$y_i^T (R_i - S_i^T \tilde{Q}_i S_i) y_i - \gamma_i \times \xi_i^- > 0, \quad \text{and} \tag{2.14}$$

$$\begin{bmatrix} \eta_i & * \\ \tilde{Q}_i S_i y_i + u_i^{ss} & I \end{bmatrix} \succcurlyeq 0; \tag{2.15}$$

Then, the QDC is control-constrained qualified w.r.t. (2.2)

Proof From (2.13), the QDC ellipsoid has the centre point at $\tilde{u}_i = -\tilde{Q}_i S_i y_i$. Its intersection with $\mathbb{U}_i$ will thus be non-empty if this centre point lies inside $\mathbb{U}_i$, i.e. if $(-\tilde{Q}_i S_i y_i - u_i^{ss})^T (-\tilde{Q}_i S_i y_i - u_i^{ss}) - \eta_i \leqslant 0$, which is equivalent to (2.15) by the Schur complement. The first condition (2.14) ensures that the aforementioned ellipsoid does not shrink to a singleton. ∎

(c) *Control qualification condition 3*

The above qualification condition may become conservative when the intersection of $\mathbb{U}_i$ and the QDC ellipsoid region governed by (2.27) is a small set. This third qualification condition alleviates such conservativeness. The following lemma provides an LMI optimisation for maximising the intersection. Define

$$\mu_i := u_i^T Q_i u_i + 2 y_i^T S_i^T u_i \quad \text{and} \quad \pi_i := \gamma_i \times \xi_i^- - y_i^T R_i y_i.$$

Lemma 2.4 *For $\pi_i > 0$, there exists $u_i \in \mathbb{U}_i$ with $u_i^{ss} = 0$, satisfying the quadratic constraint (2.13), and further, the control-constrained qualified set is maximised, if the following conditions are feasible:*

$$\mu_i^* < 0 < \pi_i,, \tag{2.16}$$

$$where \quad \mu_i^* = \text{argmin}(\mu_i) \tag{2.17}$$

$$s.t. \quad \begin{bmatrix} Q_i - \lambda_i I_{m_i} & S_i y_i \\ * & \lambda_i \times \eta_i - \mu_i \end{bmatrix} \preccurlyeq 0, \quad \lambda_i > 0.$$

Proof The LMI in (2.17) is derived from the two inequalities

$$u_i^T Q_i u_i + 2 y_i^T S_i^T u_i - \mu_i \le \lambda_i (u_i^T u_i - \eta_i) \le 0,$$

to ensure that the QDC ellipsoid region governed by (2.13) intersects the control-constraint set $\mathbb{U}_i$. Then, the minimisation (2.17) and the condition (2.16) provide the largest intersection. ■

Remark 2.3 For the qualification conditions in Lemmas 2.3 and 2.4 to be met, the time-dependent coefficient matrices $Q_i(k)$, $S_i(k)$, $R_i(k)$ may be required on case by case basis.

(d) *Control qualification condition 4*

The QDC control-constrained qualified condition can be determined as a subset of $\mathbb{U}_i$, such as the one below for the case of $u_i^{ss} = 0$:

$$\begin{bmatrix} Q_i + \lambda_i I & S_i y_i \\ * & y_i^T R_i y_i - \gamma_i \times \xi_i^- - \lambda_i \times \eta_i \end{bmatrix} \succcurlyeq 0, \quad \lambda_i \succ 0. \tag{2.18}$$

However, this subset condition may become overly conservative for many applications.

Remark 2.4 It has been identified from simulation studies that the QDC with time-dependent coefficient matrices $Q_i(k)$, $S_i(k)$, $R_i(k)$ does not always provide a better control performance than that with constant Q_i, S_i, R_i matrices. The determination of Q_i, S_i, R_i matrices, whether time-dependent or time-independent, is application specific. A universal solution will not be available.

2.2.3 Attractivity Condition for Unconstrained Systems

The asymptotic attractivity condition with the QDC is presented in this section. Firstly, define

$$\xi\big((w_i, y_i), (v_i, u_i)\big) = \begin{bmatrix} w_i \\ y_i \end{bmatrix}^T X_{\Delta i} \begin{bmatrix} w_i \\ y_i \end{bmatrix} + 2 \begin{bmatrix} w_i \\ y_i \end{bmatrix}^T Y_{\Delta i} \begin{bmatrix} v_i \\ u_i \end{bmatrix} + \begin{bmatrix} v_i \\ u_i \end{bmatrix}^T Z_{\Delta i} \begin{bmatrix} v_i \\ u_i \end{bmatrix},$$
(2.19)

whereby the coefficient matrices $X_{\Delta i}, Y_{\Delta i}, Z_{\Delta i}$ are block diagonals of the form

$$X_{\Delta i} = \text{diag}\{X_i, -Q_i\}, \ Y_{\Delta i} = \text{diag}\{Y_i, -S_i\}, \ Z_{\Delta i} = \text{diag}\{Z_i, -R_i\}, \tag{2.20}$$

in which X_i, Q_i, Z_i and R_i are symmetric matrices.

Definition 2.3 A subsystem $\mathscr{S}_i$ (2.1) is said to be *quadratically dissipative* with respect to the quadratic supply rate $\xi_i\big((w_i(k), y_i(k)), (v_i(k), u_i(k))\big)$ if there is a non-negative C^1 function $V(x_i)$, addressed as storage function, $V(x_i(0)) = 0$, such that for all pairs $\big(v_i(k), u_i(k)\big)$ and all $k \in \mathbb{Z}^+$, the following dissipation inequality is satisfied irrespectively of the initial value of the state $x_i(0)$:

$$V(x_i(k+1)) - \tau_i \times V(x_i(k)) \le \xi_i\big((w_i(k), y_i(k)), (v_i(k), u_i(k))\big), \ 0 < \tau_i \le 1. \tag{2.21}$$

The square storage function of the form $V(x_i(k)) = x_i^T(k) P_i x_i(k), \ P_i \succ 0$, is considered in this chapter. We will state and prove the asymptotic attractivity of the controlled system Σ without any control constraints in the first theorem below. The following assumption on the 'QDC attractability' must be made beforehand:

Definition 2.4 The controlled system Σ (2.4) is said to be '*QDC-attractable*' with respect to $\xi\big(u_i(k), y_i(k)\big)$, $i = 1, 2, \ldots, h$, in some neighbourhoods of the zero equilibrium $\mathbb{X} \subset \mathbb{R}^n$, $0 \in \mathbb{X}$, $\mathbb{X}$ is compact, if there exist some h control sequences $\{u_i(k) \in \mathbb{U}_i\}, i = 1, 2, \ldots, h$, such that for each $x(0) \in \mathbb{X}$, the controlled motion $(x(k), u(k))$ is negatively asymptotically attractive with respect to $\xi_{\Sigma}\big(u(k), y(k)\big)$ and $\|x(k)\| \to 0$ as $k \to \infty$ where $\xi_{\Sigma}\big(u, y\big) = \sum_1^h \xi(u_i, y_i)$.

Assumption 1 There exist some compact neighbourhoods of the zero equilibrium of Σ (2.4), $\mathbb{X} \subset \mathbb{R}^n$, $0 \in \mathbb{X}$, such that the MPC controlled system Σ with h compact control-constraint sets $\mathbb{U}_i$, $i = 1, 2, \ldots, h$, is 'QDC-attractable' in $\mathbb{X}$, with respect to $\xi\big(u_i(k), y_i(k)\big)$, $i = 1, 2, \ldots, h$.

Remark 2.5 The QDC attractability is stronger than the combined observability and controllability. The QDC attractability requires that there exist some decoupled control sequences $\{u_i(k) \in \mathbb{U}_i\}, i = 1, 2, \ldots, h$, such that the convergence of $\xi_{\Sigma}\big(u(k), y(k)\big) \to 0$ and $\|x(k)\| \to 0$ will be obtained simultaneously. When applying the QDC to the decentralised model predictive control (MPC) problem, the QDC attractability can be translated into the *recursive feasibility* [93] conditions for the local MPC optimisations at every time step.

The asymptotic attractivity condition for unconstrained systems is stated next. Define $\gamma = \max_{i=1\ldots h} \gamma_i$, $\tau = \max_{i=1\ldots h} \tau_i$. In what follows, we assume $\tau/\gamma > 1$.

Theorem 2.1 *Consider the system Σ (2.4) and a compact neighbourhood of the zero equilibrium, $\mathbb{X} \subset \mathbb{R}^n$, $0 \in \mathbb{X}$. Suppose that*

1. *The h subsystems $\mathscr{S}_i$ (2.1) of Σ, $i = 1, 2, \ldots, h$, are dissipative with respect to the supply rates $\xi_i\big((w_i, y_i), (v_i, u_i)\big)$ (2.19) for some storage functions $V(x_i(k)) = x_i^T(k) P_i x_i(k)$, $P_i \succ 0$, as per Definition 2.3, and $Q_i - S_i^T R_i^{-1} S_i \prec 0$;*

2. *The coefficient matrices X_i, Y_i, Z_i of ξ_i, $i = 1, 2, \ldots, h$, fulfil the following global condition:*

$$\mathsf{diag}[X_i]_1^h + H^T \mathsf{diag}[Z_i]_1^h H + H^T \mathsf{diag}[Y_i^T]_1^h + \mathsf{diag}[Y_i]_1^h H \prec 0; \quad (2.22)$$

3. *There are some h control sequences $\{u_i(k) \in \mathbb{R}^{m_i}\}$, $i = 1, 2, \ldots, h$, satisfying the respective quadratic constraints (2.11) for all $k > 0$, irrespectively of $x(0) \in \mathbb{X}$;*

4. *Assumption 1 holds for Σ and $\mathbb{X}$ with respect to ξ_i, $i = 1, 2, \ldots, h$, with some of the above h respective control sequences $\{u_i(k) \in \mathbb{R}^{m_i}\}$, called QDC-attractable control sequence;*

Then, the unconstrained controlled system Σ (2.4) is asymptotically attractive, i.e. $\|x\| \to 0$ as $k \to \infty$, irrespectively of $x(0) \in \mathbb{X} \subset \mathbb{R}^n$.

Proof Using (2.21) and $v = Hw$, we obtain for $V(x(k)) = \sum_{i=1}^h V(x_i(k))$,

$$V(x(k+1)) - \tau \times V(x(k)) \leqslant \sigma(k), \quad 0 < \tau < 1,$$

where $\sigma(k) := \sum_{i=1}^h \xi\big((w_i(k), y_i(k)), (v_i(k), u_i(k))\big)$

$$= w^T(k)\Big[\mathsf{diag}[X_i]_1^h + H^T \mathsf{diag}[Z_i]_1^h H + 2\mathsf{diag}[Y_i]_1^h H\Big]w(k)$$

$$- \sum_{i=1}^h [y_i^T(k) \ u_i^T(k)] \begin{bmatrix} R_i & S_i^T \\ S_i & Q_i \end{bmatrix} \begin{bmatrix} y_i(k) \\ u_i(k) \end{bmatrix},$$

which yields

$$V(x(k+1)) \leqslant \tau \times V(x(k)) + \gamma |\xi^-|,$$

due to (2.10) and the two conditions (2) and (3) of this theorem.

Then, by iteration we obtain for an initial time $k_0 \geq 0$,

$$V(x(k)) \leqslant \tau^{k-k_0} \times V(x(k_0)) + \tau^{k-k_0} \times |\xi(k_0 - 1)| \times (1 + \varepsilon + \cdots + \varepsilon^{k-k_0-1}),$$

$$= \tau^{k-k_0} \times V(x(k_0)) + \tau^{k-k_0} \times |\xi(k_0 - 1)| \times \frac{1 - \varepsilon^{k-k_0}}{1 - \varepsilon}, \quad \varepsilon = \gamma/\tau < 1,$$

where $\xi(k_0) := \xi\big(u(k_0), y(k_0)\big)$.

For the convergence of $V(x(k))$, it is to prove that for each $\beta > 0$ there is a time instant $k(\beta) > k_0$ such that $V(x(k)) \leqslant \beta \ \forall k \geqslant k(\beta)$.

Indeed, for each $\beta \geqslant 0$, there exist two time instants $\bar{k} > k_0$ and $\tilde{k} > k_0$ such that

$$\gamma^{\bar{k}-k_0} \times |\xi(k_0 - 1)| \leqslant \frac{\beta}{2}, \text{ and } \tau^{\tilde{k}-k_0} \times V\big(x(k_0)\big) \leqslant \frac{\beta}{2},$$

since $0 < \gamma < 1$ and $0 < \tau < 1$. Therefore, there exists $\check{k} = \sup(\bar{k}, \tilde{k})$ such that for each $\beta \geq 0$,

$$V\big(x(k)\big) < \frac{\beta}{2} + \frac{\beta}{2} = \beta \ \forall k \geqslant \check{k}, \text{ due to } \varepsilon < 1.$$

With $P \succ 0$, we have $\lambda_{\min}(P)x^2 \leqslant V\big(x(k)\big) \leqslant \lambda_{\max}(P)x^2$; therefore, $\|x(k)\| \to 0$ as $k \to +\infty$, if $\lambda_{\min}(P)$ and $\lambda_{\max}(P)$ are bounded, which is the case with a full row rank P. The proof for state convergence is complete.

Next, the feasibility of the QDC at every time step $k > k_0$ must be obtained, irrespectively of the initial $x(k_0)$, for the asymptotic attractivity to be guaranteed. This is a direct result of Assumption 1 for the unconstrained system Σ and the two conditions (1), (3) and (4) of this theorem. The proof is thus complete. ∎

The proof for the attractivity condition without assuming $\tau/\gamma > 1$ is provided in Appendix A.2. The next proposition states the asymptotic attractivity condition in Theorem 2.1 with LMI dissipative criteria.

Proposition 2.1 *Consider the system Σ (2.4) and a compact neighbourhood of the zero equilibrium $\mathbb{X} \subset \mathbb{R}^n$, $0 \in \mathbb{X}$. Suppose that*

1. *The coefficient matrices Q_i, S_i, R_i, X_i, Y_i, Z_i of $\xi\big((w_i, y_i), (v_i, u_i)\big)$, $i = 1, 2, \ldots, h$, (2.19) satisfy the following conditions:*

   a. *Local condition:*

$$\begin{bmatrix} A_i^T P_i A_i - \tau_i P_i + C_i^T R_i C_i - F_i^T X_i F_i & A_i^T P_i B_i + C_i^T S_i & A_i^T P_i E_i - F_i^T Y_i \\ * & B_i^T P_i B_i + Q_i & B_i^T P_i E_i \\ * & * & E_i^T P_i E_i - Z_i \end{bmatrix} \prec 0, \tag{2.23}$$

$$\begin{bmatrix} Q_i & S_i \\ * & R_i \end{bmatrix} \prec 0, \quad P_i \succ 0, \tag{2.24}$$

$$i = 1, 2, \ldots h,$$

   b. *Global condition:*

$$\mathsf{diag}[X_i]_1^h + H^T \mathsf{diag}[Z_i]_1^h H + H^T \mathsf{diag}[Y_i^T]_1^h + \mathsf{diag}[Y_i]_1^h H \prec 0; \tag{2.25}$$

2. *There are some QDC-attractable control sequences $\{u_i(k) \in \mathbb{R}^{m_i}\}$, $i = 1, 2, \ldots, h$, satisfying the respective quadratic constraints (2.11) for all $k > 0$, irrespectively of $x(0) \in \mathbb{X}$;*

Then, the unconstrained controlled system Σ (2.4) is asymptotically attractive.

Proof This is obtained from Theorem 2.1 by deriving the LMI dissipative criterion from the dissipation inequality (2.23) for each single subsystem $\mathscr{S}_i$, see, e.g. [17]. ■

The LMI

$$Q_i - S_i^T R_i^{-1} S_i \prec 0$$

in the condition (1) of Theorem 2.1 leads to a non-positive supply rate ξ_i for all $k \in \mathbb{Z}$. The QDC is thus applicable in Theorem 2.1. The following theorem states the attractivity condition with the APRC:

Theorem 2.2 *Consider the system Σ (2.4) and a compact neighbourhood of the zero equilibrium, $\mathbb{X} \subset \mathbb{R}^n$, $0 \in \mathbb{X}$. Suppose that*

1. *The h subsystems $\mathscr{S}_i$ (2.1) of Σ, $i = 1, 2, \ldots, h$, are dissipative with respect to the supply rates $\xi_i((w_i, y_i), (v_i, u_i))$ (2.19) with the respective storage functions $V(x_i(k)) = x_i^T(k) P_i x_i(k)$, $P_i \succ 0$, as per Definition 2.3;*

2. *The coefficient matrices X_i, Y_i, Z_i of ξ_i, $i = 1, 2, \ldots, h$, fulfil the following global condition:*

$$\text{diag}[X_i]_1^h + H^T \text{diag}[Z_i]_1^h H + H^T \text{diag}[Y_i^T]_1^h + \text{diag}[Y_i]_1^h H \prec 0; \quad (2.26)$$

3. *The initial supply rates $\xi_i(u_i(0), y_i(0)) < 0$ for $i = 1, 2, \ldots, h$;*
4. *There are some QDC-attractable control sequences $\{u_i(k) \in \mathbb{R}^{m_i}\}$, $i = 1, 2, \ldots, h$, satisfying the respective quadratic constraints in the following for all $k > 0$, irrespectively of $x(0) \in \mathbb{X}$:*

$$u_i(k)^T Q_i u_i + 2 y_i^T S_i^T u_i(k) \geqslant \gamma_i \times \xi_i^- - y_i^T R_i y_i + \mu_i(k-1) - \mu_i(k);$$
$$(2.27)$$

Then, the unconstrained controlled system Σ (2.4) is asymptotically attractive, i.e. $\|x\| \to 0$ as $k \to \infty$, irrespectively of $x(0) \in \mathbb{X} \subset \mathbb{R}^n$.

Proof The proof is similar to that for Theorem 2.1 except that the global supply rate $\xi = \sum_{i=1}^h \xi_i$ will turn to positive after the time step κ_s that is the maximum of the time steps k_s corresponding to ξ_i, $i = 1, 2 \ldots, h$, i.e. $\xi > 0 \; \forall k \geqslant \kappa_s$. Then we will obtain from the dissipation inequality that $V(x(k)) \leqslant \tau V(x(k-1))$. The state convergence is obtained accordingly. ■

Before we state the attractivity conditions for the constrained Σ, the decentralised MPC with the QDC is formulated in the next section.

2.3 Decentralised Model Predictive Control and Quadratic Constraint

2.3.1 Decentralised Model Predictive Control

Consider a conventional MPC cost function with respect to the state and control input vectors [88], and adequately chosen weighting matrices $\mathcal{Q}_i$, $\mathcal{R}_i$ and a predictive horizon N, for a subsystem $\mathcal{S}_i|$ stand alone (2.5), as follows:

$$\mathcal{J}_i^k = \sum_{\ell=1}^{N+1} x_i^T(k+\ell)\mathcal{Q}_i x_i(k+\ell) + \sum_{\ell=0}^{N} u_i^T(k+\ell)\mathcal{R}_i u_i(k+\ell). \qquad (2.28)$$

The constrained optimisation problem of a local MPC at the time step k is then as follows:

$$\min_{\hat{u}_i(k)} \quad \hat{u}_i^T(k)\Phi_i \hat{u}_i(k) + 2\Upsilon_i(k)\hat{u}_i(k) + \delta_i(k) \quad \text{subject to} \quad \hat{u}_i(k) \in \hat{\mathbb{U}}_i, \qquad (2.29)$$

where $\Phi_i := \Gamma_i^T \tilde{\mathcal{Q}}_{iN} \Gamma_i + \tilde{\mathcal{R}}_{iN}$, $\Upsilon_i(k) := \mathrm{r}_i^T(k)\tilde{\mathcal{Q}}_{iN}\Gamma_i$, $\delta_i(k) := \mathrm{r}_i^T(k)\tilde{\mathcal{Q}}_{iN}\mathrm{r}_i(k)$,

$$\mathrm{r}_i(k) := \Theta_i x_i(k), \quad \tilde{\mathcal{Q}}_{iN} := \mathrm{diag}[\mathcal{Q}_i]_1^{N_i}, \quad \tilde{\mathcal{R}}_{iN} := \mathrm{diag}[\mathcal{R}_i]_1^{N_i},$$

$$\Gamma_i := \begin{bmatrix} B_i & \ldots & 0 & 0 \\ A_i B_i & \ldots & 0 & 0 \\ \ldots & \ldots & \ldots & \ldots \\ A_i^N B_i & \ldots & A_i B_i & B_i \end{bmatrix}, \quad \Theta_i := \begin{bmatrix} A_i \\ A_i^2 \\ \ldots \\ A_i^{N+1} \end{bmatrix},$$

and $\hat{\mathbb{U}}_i$ is the constraint set for the variable $\hat{u}_i$, which is deduced from the constraint set $\mathbb{U}_i$ (2.21), specifically,

$$\hat{\mathbb{U}}_i := \{\hat{u}_i = [u_i^T(k), u_i^T(k+1), \ldots, u_i^T(k+N)]^T : u_i \in \mathbb{U}_i\}. \qquad (2.30)$$

Using the QDC-based attractivity constraint (2.27), the local MPC optimisation is cast as a convex quadratically constrained quadratic programming (QCQP) problem of the form:

$$\min_{\hat{u}_i(k)} \quad \hat{u}_i(k)^T \Phi_i \hat{u}_i(k) + 2\Upsilon_i(k)\hat{u}_i(k) + \delta_i(k) \qquad (2.31)$$

$$\text{s.t.} \quad \hat{u}_i(k)^T M_i^T Q_i M_i \hat{u}_i(k) + 2y_i^T S_i M_i \hat{u}_i(k) + \pi_i \geqslant 0, \qquad (2.32)$$

$$\text{and} \quad \hat{u}_i(k) \in \hat{\mathbb{U}}_i, \qquad (2.33)$$

where $M_i = [I_{m_i} \, 0_{m_i} \, \ldots \, 0_{m_i}]$, $\pi_i := \gamma_i \times \xi_i^- + y_i^T R_i y_i$.

The QCQP problem (2.31) is then solved by the local solver for the minimising vector sequence $\hat{\boldsymbol{u}}_i^*(k)$ which consists of N elements of $u_i^*(k + \ell)$, $\ell = 0, 1, \ldots, N$. Only the first element $u_i^*(k)$ of the sequence $\hat{\boldsymbol{u}}_i^*(k)$ is output to $\mathscr{S}_i$. This rolling process is repeated at the next time step and continues thereon. For the convex optimisation problem (2.31), we assume without loss of generality that $\hat{\boldsymbol{u}}_i^*(k)$ is uniquely defined herein.

Remark 2.6 The optimisation (2.31) may become occasionally infeasible in some applications, such as the first example in Sect. 2.5. We can add a small negative number to the right-hand side of the quadratic constraint (2.32) to eliminate such locally infeasible issue.

Remark 2.7 The optimisation (2.31) may be infeasible if one of the qualification conditions in Sect. 2.2.2 does not hold true.

Remark 2.8 The optimisation (2.31) will be modified as follows when the APRC is used instead of the QDC, assuming $\xi_i\big(u_i(0), y_i(0)\big) < 0$ and $\xi_i \geqslant 0 \ \forall k > k_s$:

$$\min_{\hat{u}_i(k)} \quad (2.31) \tag{2.34}$$

$$\text{s.t.} \quad (2.32) \ \text{if} \ \xi_i^{-1} < 0, \quad \text{and} \quad (2.33). \tag{2.35}$$

The MPC optimisation (2.34) will not, however, guarantee that the supply rate ξ_i will always be positive once the quadratic constraint (2.32) is not included. Therefore, the supply rate can become negative again in some future time. The boundedness condition that allows for several repeated time instants k_s will be stated in a proposition below. We firstly define the repeatedly energy-dissipative constraint in the sequel. Denote the set of integers in the interval $[i, j]$ as $\mathbb{I}_{[i,j]}$, $j > i$.

Definition 2.5 The controlled motion $\big(y(k), u(k)\big)$ of Σ (2.4) is said to satisfy the *repeatedly energy-dissipative constraint* (REDC) w.r.t $\xi(u, y)$ if, for $0 < \gamma_\sigma < 1$, the initial supply rate $\xi\big(u(0), y(0)\big) < 0$, the following inequalities hold:

$$\begin{cases} \xi\big(u(k), y(k)\big) \geqslant \gamma_\sigma \times \xi\big(u(k-1), y(k-1)\big) + \mu(k), \ \forall k \in \mathbb{I}_{[\kappa_\sigma, \kappa_{\sigma+1})}, \\[2mm] \text{and} \ \ \xi\big(u(\kappa_{\sigma+2}), y(\kappa_{\sigma+2})\big) < \xi\big(u(\kappa_{\sigma+1}), y(\kappa_{\sigma+1})\big) < 0, \end{cases}$$

$$(2.36)$$

$$\text{for all} \ \mathbb{I}_{[\kappa_\sigma, \kappa_{\sigma+1})} \subset \bigcup_{\sigma \in \mathbb{J} \subset \mathbb{Z}_0^+} \mathbb{I}_{[\kappa_\sigma, \kappa_{\sigma+1})} := \mathbb{I}_{[0, \kappa_1)} \cup \mathbb{I}_{[\kappa_1, \kappa_2)} \cup \mathbb{I}_{[\kappa_3, \kappa_4)} \cup \ldots.$$

The Lagrange uniform boundedness stated in the next proposition has been defined in [96].

Proposition 2.2 *Consider the system Σ (2.4) and a compact neighbourhood of the zero equilibrium, $\mathbb{X} \subset \mathbb{R}^n$, $0 \in \mathbb{X}$. Suppose that*

Fig. 2.2 Illustrations for the repeatedly energy-dissipative constraint (REDC)

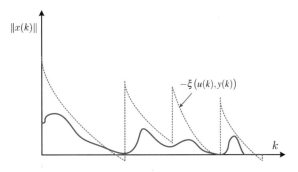

1. *The controlled system Σ is dissipative with respect to $-\xi\big(u(k), y(k)\big)$, with the following dissipation inequality:*

$$V(x(k+1)) - \gamma \times V(x(k)) \leqslant -\xi\big(u(k), y(k)\big), \quad V(x) = x^T P x, \quad P \succ 0;$$
(2.37)

2. *The initial supply rate is negative, i.e. $\xi\big(u(0), y(0)\big) < 0$;*
3. *There are some h control sequences $\{u_i(k) \in \mathbb{R}^{m_i}\}$, $i = 1, 2, \ldots, h$, such that the REDC (2.36) holds for all $k > 0$, irrespectively of $x(0) \in \mathbb{X}$;*

Then, the unconstrained controlled system Σ (2.4) is Lagrange uniformly bounded.

The representation for the REDC and state boundedness is depicted by Fig. 2.2. The supply rate is monotonically decreasing in some time intervals and may become positive, but repeatedly restarts the decreasing process from the negative domain.

The state estimation for output feedback problems is outlined in the next section.

2.3.2 Centralised Moving Horizon State Estimation

For an output feedback problem, consider the following model:

$$\Sigma_0 : \begin{cases} x(k+1) = Ax(k) + Bu(k) + d(k), \\ z(k) = C'x(k) + e(k), \end{cases}$$
(2.38)

where $x \in \mathbb{R}^n$ and $u \in \mathbb{U} \subset \mathbb{R}^m$ are the state and control vector, respectively; $d \in \mathbb{W} \subset \mathbb{R}^n$ is the persistent, unknown but bounded, disturbance vector; $z \in \mathbb{R}^q$ is the measurement output vector; and $e \in \mathbb{V} \subset \mathbb{R}^q$ is the measurement error vector.

The state vector $x(k)$ is estimated from the measurement output $z(k)$ and the historical control $u(.)$ before being used in the MPC optimisation. The moving horizon estimation (MHE) algorithm developed by the authors in [32, 123] can be adopted for this state estimation problem for the constrained Σ_0. Denote the estimated state as $\hat{x}$ and the estimated disturbance sequence

$$\{\hat{d}(k)\} := \{\hat{d}(k - N + 1), \hat{d}(k - N + 2), \dots, \hat{d}(k - 1)\}.$$

At every time step, the state $x(k)$ is estimated by formulating and solving the MHE optimisation with the following objective function:

$$\Psi(\hat{x}(k - N + 1), \{\hat{d}(k)\}) = \phi(k - N + 1) + \mathscr{J}(k, N, z(k)),$$

where the main cost function is defined as

$$\mathscr{J}(k, N, z(k)) := \sum_{\ell=k-N+1}^{k} \|z(\ell) - C'x(\ell)\|^2_{\mathscr{R}^{-1}} + \sum_{\ell=k-N+1}^{k-1} \|d(\ell)\|^2_{\mathscr{Q}^{-1}},$$

and the *arrival cost function* $\phi(k - N + 1)$ is given as

$$\phi(k - N + 1) = \frac{1}{2}\|x(k - N + 1) - \hat{x}(k - N + 1|k - N)\|^2_{\mathscr{P}^{-1}_{k-N+1|k-N}}, \quad (2.39)$$

in which

$$\begin{aligned}
\mathscr{P}_{k-N+1|k-N} = {}& A\mathscr{P}_{t-N|t-N-1}A^T - A\mathscr{P}_{k-N|k-N-1}C'^T \quad (2.40)\\
&\times (\mathscr{R} + C'\mathscr{P}_{k-N|k-N-1}C'^T)^{-1}C'\mathscr{P}_{k-N|k-N-1}A^T + \mathscr{Q}.
\end{aligned}$$

The MHE optimisation problem is summarised as follows:

$$\Theta(k)^* = \min_{x(k-N+1),\{\hat{d}(k)\}} \Psi(x(k - N + 1), \{\hat{d}(k)\}) \quad (2.41)$$

subject to

$$\begin{aligned}
&x(\ell) = Ax(\ell - 1) + Bu(\ell - 1) + d(\ell - 1), \text{ for } \ell = k - N + 1, \dots, k\\
&z(\ell - 1) = C'x(\ell - 1) + e(\ell - 1),\\
&x(\ell) \in \mathbb{X}, \ d(\ell) \in \mathbb{W}, \ e(\ell) \in \mathbb{V}.
\end{aligned}$$

The estimated state is then obtained by the following prediction model:

$$\hat{x}(k) = A^{N-1}\hat{x}(k - N + 1) + \sum_{j=k-N+1}^{k-1} A^{k-j-1}[Bu(j) + \hat{d}(j)]. \quad (2.42)$$

The weighting matrices $\mathscr{Q}^{-1}$ and $\mathscr{R}^{-1}$ are usually chosen as the variances of the respective variables. Once the estimated state $x(k) = \hat{x}(k)$ is computed, the local MPC optimisations (2.31) are then formulated for h subsystems and solved for the global minimising sequence $\hat{\mathbf{u}}_i^*(k)$. And only the first element $u_i^*(k)$ is output to control Σ_0 (2.38).

2.3.3 Attractivity Condition for Control-Constrained Systems

The following theorem states the asymptotic attractivity condition for the control-constrained Σ, which is applicable to the decentralised MPC in Sect. 2.3.1.

Theorem 2.3 *Consider the system Σ (2.4) and a compact neighbourhood of the zero equilibrium, $\mathbb{X} \subset \mathbb{R}^n$, $0 \in \mathbb{X}$, and $x(0) \in \mathbb{X}$. Suppose that*

1. *The h subsystems $\mathscr{S}_i$ (2.1) of Σ, $i = 1, 2, \ldots, h$, are dissipative with respect to the supply rates $\xi_i((w_i, y_i), (v_i, u_i))$ (2.19) with the respective storage functions $V(x_i(k)) = x_i^T(k) P_i x_i(k)$, $P_i \succ 0$, as per Definition 2.3, and $Q_i - S_i^T R_i^{-1} S_i \prec 0$;*

2. *The coefficient matrices X_i, Y_i, Z_i of ξ_i, $i = 1, 2, \ldots, h$, fulfil the following global condition:*

$$\mathrm{diag}[X_i]_1^h + H^T \mathrm{diag}[Z_i]_1^h H + H^T \mathrm{diag}[Y_i^T]_1^h + \mathrm{diag}[Y_i]_1^h H \prec 0; \quad (2.43)$$

3. *One of the qualification conditions in Sect. 2.2.2 holds true for all $k > 0$;*

4. *The convergence rates of $\xi_i(y_i(k), u_i(k)) \to 0$, $i = 1, 2, \ldots, h$, are sufficiently fast such that the respective local MPC optimisations (2.31) are recursively feasible at every time step $k > 0$;*

Then, the constrained controlled system Σ (2.4) is asymptotically attractive.

Proof This is a direct result of Theorem 2.1 with the QDC attractability obtained from the recursive feasibilities at every time step of all local MPC optimisations (2.31) by fulfilling the condition (4), plus the QDC control-constrained qualification conditions in Sect. 2.2.2 as required by the condition (3). ∎

Remark 2.9 The last conditions (4) in Theorem 2.3 will be met when a global terminal condition of the form $x(k + N) \in \Omega$, in which Ω is the *maximal output admissible set* defined in [73], is satisfied at every time step, which will not be the case if all the supply rates $\xi_i(y_i(k), u_i(k))$, $i = 1, 2, \ldots, h$, will not converge sufficiently fast. This is due to the fact that the recursive feasibility of a local MPC may not be guaranteed in advance without having the global terminal condition $x(k + N) \in \Omega$ to be fulfilled at every time step, and consequently, the condition (4) will not be met because $x(k + 1) \in \mathbb{R}^n/\mathbb{X}$, instead of $x(k + 1) \in \mathbb{X}$. This will be elaborated further in Remark 2.10.

The decentralised MPC algorithm with the quadratic constraint is then provided in the next section.

2.4 Decentralised MPC with Quadratic Constraint Algorithm

2.4.1 Procedure

The pseudo-algorithm of a decentralised model predictive control with the quadratic constraint is outlined in Procedures 2.1 and 2.2 below.

Procedure 2.1 *Off-line Computation*:

1. Initiate ξ_i^0, $y_i(0)$, $\Upsilon_i(0)$, $\delta_i(0)$, $x_i(1)$ and select the coefficient γ_i.
2. Determine the Q_i, S_i, R_i coefficient matrices of the QDC from a feasible solution to (2.23), (2.24) and (2.25).
3. Calculate Φ_i.

Procedure 2.2 *Online Computation*: At every time step $k \geq k_0$,

Each local subsystem $\mathscr{S}_i$ will

1. Update $\Upsilon_i(k)$ and $\delta_i(k)$;
2. Solve the optimisation problem (2.31) to yield the local control sequence $\hat{u}_i^*(k)$.
3. The first optimising vector $u_i^*(k)$ is applied to control $\mathscr{S}_i$;
4. Return to 1.

The LMIs in (2.23)–(2.25) will be assuredly feasible with the LMI optimisation problems in the next subsection.

2.4.2 Determination of the QDC Coefficient Matrices

The coefficient matrices Q_i, S_i, R_i, X_i, Y_i, Z_i in Proposition 2.1 and Theorem 2.3, and in step (2) of Procedure 2.1 above, can be determined from the following LMI optimisation problem:

$$\min_{X_{\Delta i},\, Y_{\Delta i},\, Z_{\Delta i}} \quad x^T(0)Q_o x(0) \tag{2.44}$$

$$\text{subject to} \quad (2.23),\ (2.24),\ \text{and}\ (2.25),$$

where $Q_o := \text{diag}[Q_i]_1^h$, if $x(0) \neq 0$.

Alternatively, the following global dissipative condition can be employed in the LMI optimisation for determining the Q_i, S_i, R_i coefficient matrices of the QDC, as follows:

$$\min_{Q_i,\ S_i,\ R_i} \quad x^T(0)Q_o x(0) \tag{2.45}$$

$$\text{subject to} \quad \begin{bmatrix} A^T PA - \tau P + C^T Q_o C & A^T PB + C^T S_o \\ * & B^T PB + R_o \end{bmatrix}, \ P \succ 0, \tag{2.46}$$

$$\begin{bmatrix} Q_i & S_i \\ * & R_i \end{bmatrix} \prec 0, \ i = 1, \dots, h, \tag{2.47}$$

where $Q_o := \mathsf{diag}[Q_i]_1^h$, $S_o := \mathsf{diag}[S_i]_1^h$, $R_o := \mathsf{diag}[R_i]_1^h$, if $x(0) \neq 0$.

The proposition that states the asymptotic attractivity condition with these LMI optimisations is given below.

Proposition 2.3 *Consider the system Σ (2.4) and a compact neighbourhood of the zero equilibrium, $\mathbb{X} \subset \mathbb{R}^n$, $0 \in \mathbb{X}$, and $x(0) \in \mathbb{X}$. Suppose that*

1. *The Q_i, S_i, R_i coefficient matrices of the QDCs of h subsystems $\mathscr{S}_i$, $i = 1, 2, \dots, h$, are the solution to the LMI optimisation of either (2.44) or (2.45);*

2. *One of the qualification conditions in Sect. 2.2.2 holds true for all $k > 0$;*

3. *The convergence rates of $\xi_i\big(y_i(k), u_i(k)\big) \to 0$, $i = 1, 2, \dots, h$, are sufficiently fast such that the respective local MPC optimisations (2.31) are recursively feasible at every time step $k > 0$;*

Then, the constrained controlled system Σ (2.4) is asymptotically attractive.

Proof This is a direct result of Proposition 2.1 and Theorem 2.3, and the employment of block-diagonal structures of Q_o, S_o, R_o matrices in (2.45) and (2.44). The asymptotic attractivity of the global system Σ is obtained by applying the LMI dissipative criterion for the global system Σ (2.46), instead of the dissipative criteria for h subsystems $\mathscr{S}_i$ (2.23) and the global condition (2.25) as in Proposition 2.1. The proof for the asymptotic attractivity of Σ in Theorem 2.1 then remains unchanged when the X_i, Y_i, Z_i matrices vanish, and the global storage function $V(x(k))$ is chosen instead of $\mathsf{diag}[V_i(x_i(k))]_1^h$. ∎

Remark 2.10 For the last condition in Proposition 2.3 to be fulfilled by a global terminal condition as mentioned in Remark 2.9, an additional constraint in $u_i(k)$ to be derived from the 'global terminal constraint set' together with the local 'one-step admissible control sets' of respective subsystems $\mathscr{S}_i$, $i = 1, 2, \dots, h$, will have to be imposed onto each local MPC optimisation (2.31); see, e.g. [160]. This is not under the scope of this chapter, nevertheless, but only an assumption that the solution to (2.31) deems hold the condition (3) true and are always recursively feasible at every time step. This also means that all the local supply rates will converge to zero sufficiently fast such that the convergence of the global state, $x(k) \to 0$, will be achievable by the solutions to the local optimisations in the next following time steps, and the local MPC optimisations will thus be recursively feasible guaranteed as a result. The recursive feasibility for a local MPC problem will be further discussed in the next chapter, Appendix A.

The illustrative examples with a helicopter model and two networks of process systems are given in the next section to demonstrate the effectiveness of the decentralised MPC with quadratic constraints.

2.5 Numerical Simulation

Numerical simulation has been studied with MATLAB robust control and MPC toolboxes, as well as with Yalmip toolbox and SeDuMi v1.3 SDP solver. The QDC is employed in the first example while the APRC is applied in the other two examples.

2.5.1 Illustrative Example 1

A helicopter model of

$$
A = \begin{bmatrix} -0.02 & 0.005 & 2.4 & -32 \\ -0.14 & 0.44 & -1.3 & -30 \\ 0 & 0.018 & -1.6 & 1.2 \\ 0 & 0 & 1.0 & 0 \end{bmatrix}, \quad B = \begin{bmatrix} 0.14 & -0.12 \\ 0.36 & -8.6 \\ 0.35 & 0.009 \\ 0 & 0 \end{bmatrix}, \quad C = I,
$$

has been simulated with a centralised MPC with the QDC-based attractivity constraint as in (2.31), but for the system Σ instead of $\mathscr{S}_i$. This is a CH-47 tandem-rotor helicopter in horizontal motion as discussed in [30]. The Yalmip toolbox with SeDuMi v1.3 solver has been used in this numerical example.

The system is discretised with a sampling time of $T_s = 0.25$. The initial state vector is set at $[0.5\ 0.3\ -0.4\ 0.3]^T$. The MPC cost function has the following weighting coefficients: $\mathscr{Q} = \mathsf{diag}[0.02,\ 0.02]$, $\mathscr{R} = \mathsf{diag}[1,\ 1,\ 2,\ 2]$. The controlled system is unstable with the MPC predictive horizon of $N = 3$ when the quadratic attractivity constraint is absent. When applying the QDC-based attractivity constraint to the MPC optimisation problem (2.31) for the global system Σ, the system is asymptotically attractive as shown in Fig. 2.3. In this simulation study, the matrix P of the storage function $V(x) = x^T P x$ has not been the decision variable in the LMI optimisation (1), but determined by the Riccati equation in the LQR design. The QDC-based attractivity constraint has been switched off from the MPC optimisation problem (2.31) whenever the supply rate is larger than -25, i.e. $-\xi < 25$, in this simulation study to avoid the numerical error. During the simulation time in Fig. 2.3, the QCQP with the attractivity constraint has been occasionally infeasible, and the attractivity constraint has been ignored in those incidences.

The off-line computed coefficient matrices of the QDC has been employed in this simulation study. These coefficient matrices and the rates of changes of $V(x(k))$ and $\xi(u(k), x(k))$ along the trajectories will have to be tuned. There are three operational regions for tuning indicated in Fig. 2.4, as follows:

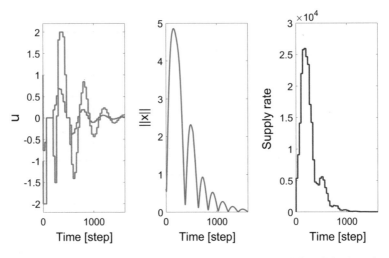

Fig. 2.3 Centralised MPC with quadratic attractivity constraint. From left to right, the trajectories of $u(k)$, $\|x(k)\|$ and $-\xi(u(k), x(k))$

Fig. 2.4 Quadratic constraint tuning regions—N is the MPC predictive horizon and Ts is the sampling rate

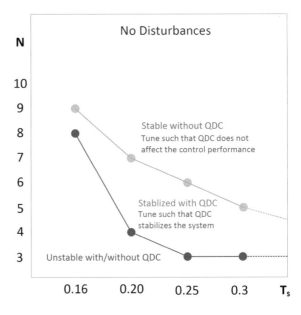

- The region above the green line is the stable region, in which the controlled system is stable without the quadratic constraint. The tuning is such that the quadratic constraint does not affect the control performance since the quadratic constraint is integrated into the MPC optimisation problem.
- The region beneath the red line is the unstable region, in which the controlled system is unstable, and the quadratic constraint cannot stabilise the system.

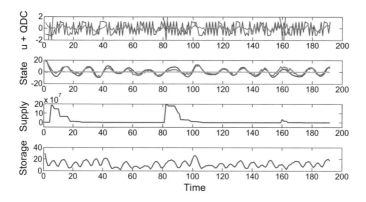

Fig. 2.5 Centralised MPC with REDC-based quadratic constraint. From top to bottom, the trajectories of $u(k)$, $x_i(k)$ ($i = 1, 2$), $-\xi(u(k), x(k))$ and $V(x(k))$

- The region in the middle is the quadratic constraint stabilisable region, in which the system is unstable without the quadratic constraint, and can be stabilised by a quadratic constraint. The tuning is required to achieve the stabilisation.

The controlled system is also unstable with the sampling time of $T_s = 0.675$ and the MPC predictive horizon of $N = 3$. When applying the REDC-based quadratic constraint to the MPC, the system is uniformly bounded with multiple asymptotic gains (the supply rate surges to a large negative value before decreasing several times) as shown in Fig. 2.5. The REDC is only applied in some interval as shown by the red colour line. There are three asymptotic gains in this simulation study corresponding to three infeasible incidences occurring from the QCQP optimisation solver.

2.5.2 Illustrative Example 2

The following state-space models of three subsystems and the global coupling matrix are considered in this numerical example:

$$\dot{x}(t) = \begin{bmatrix} A_1 & & \\ & A_2 & \\ & & A_3 \end{bmatrix} x(t) + \begin{bmatrix} B_1 & & \\ & B_2 & \\ & & B_3 \end{bmatrix} u(t) + \begin{bmatrix} E_1 & & \\ & E_2 & \\ & & E_3 \end{bmatrix} v(t), \quad (2.48)$$

$$w(t) = \begin{bmatrix} F_1 & & \\ & F_2 & \\ & & F_3 \end{bmatrix} x(t), \quad v(t) = Hw(t), \quad (2.49)$$

$$y(t) = \begin{bmatrix} C_1 & & \\ & C_2 & \\ & & C_3 \end{bmatrix} u(t), \quad (2.50)$$

where

$$A_1 = \begin{bmatrix} -1.4 & 0.3 & 0 \\ 0 & -1.8 & 1.5 \\ 0.1 & -2.7 & 1.06 \end{bmatrix}, B_1 = \begin{bmatrix} 0 \\ 0 \\ 1 \end{bmatrix}, E_1 = \begin{bmatrix} 5.1 \\ 0 \\ 3.8 \end{bmatrix}, C_1 \begin{bmatrix} 1 & 0 & 0 \\ 0 & 1 & 0 \\ 0 & 0 & 1 \end{bmatrix}, F_1 = \begin{bmatrix} 1 & 0 & 0 \end{bmatrix},$$

$$A_2 = -\begin{bmatrix} 0.76 & 0 & -.25 \\ .48 & -.56 & 0 \\ 0 & -.2 & 0.34 \end{bmatrix}, B_2 = -\begin{bmatrix} 1 & 0 & 0 \\ 0 & 1 & 0 \\ 0 & 0 & 1 \end{bmatrix}, E_2 = \begin{bmatrix} 4.2 \\ 0 \\ 0 \end{bmatrix}, C_2 = \begin{bmatrix} 1 & 0 & 0 \\ 0 & 1 & 0 \\ 0 & 0 & 1 \end{bmatrix}, F_2 = \begin{bmatrix} 0 & 0 & 1 \end{bmatrix},$$

$$A_3 = \begin{bmatrix} -4.3 & 5.9 \\ -1.8 & 2.7 \end{bmatrix}, B_3 = \begin{bmatrix} 1 & 0 \\ 0 & 1 \end{bmatrix}, E_3 = \begin{bmatrix} 6.95 \\ 0 \end{bmatrix}, C_3 = \begin{bmatrix} 1 & 0 \\ 0 & 1 \end{bmatrix}, F_3 = \begin{bmatrix} 0 & 1 & 0 \end{bmatrix}, H = \begin{bmatrix} 0 & 0 & 1 \\ 1 & 0 & 0 \\ 0 & 0 & 0 \end{bmatrix}.$$

The predictive horizons $N_i \equiv 6$, $i = 1, 2, 3$, have been chosen for all subsystems. The MPC weighting matrices are as follows:

$$\mathcal{R}_1 = \mathsf{diag}\{0.2, 0.2\}, \quad \mathcal{Q}_1 = \mathsf{diag}\{1, 1, 2\}, \quad \mathcal{R}_2 = \mathsf{diag}\{0.2, 0.2, 0.2\},$$
$$\mathcal{Q}_2 = \mathsf{diag}\{1.5, 2.4, 1.6\}, \quad \mathcal{R}_3 = \mathsf{diag}\{0.5, 0.5\}, \quad \mathcal{Q}_3 = \mathsf{diag}\{1, 3\}.$$

The constraints on the control inputs are set up with

$$\|u_1\| \leqslant 2, \quad \|u_2\| \leqslant 2, \quad \|u_3\| \leqslant 2.$$

The initial states are set at

$$x_1(0) = \begin{bmatrix} 8 \\ -7 \\ -8 \end{bmatrix}, \quad x_2(0) = \begin{bmatrix} -6 \\ 8 \\ -1.5 \end{bmatrix}, \quad x_3(0) = \begin{bmatrix} .55 \\ -1.4 \end{bmatrix}.$$

The decentralised MPC without the APRC for the unconstrained system is not able to stabilise the interconnected system, as shown in Fig. 2.6. This system is open-loop unstable with the eigenvalues of A as follows:

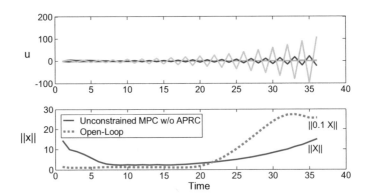

Fig. 2.6 Decentralised MPCs without both APRC stability and control constraints

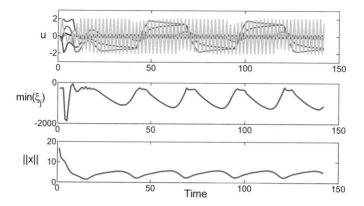

Fig. 2.7 Decentralised MPCs without APRC stability constraints

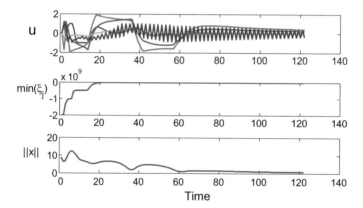

Fig. 2.8 Decentralised MPCs with APRCs and re-computed coefficient matrices

$$1.1634 + 0.2051i, \ 1.1634 - 0.2051i, \ 0.5411 + 0.1302i,$$
$$0.5411 - 0.1302i, \ 0.8628 + 0.3398i, \ 0.8628 - 0.3398i,$$
$$0.7623 + 0.3412i \ \text{and} \ 0.7623 - 0.3412i.$$

Figure 2.7 depicts the plant output time responses with the decentralised MPC without the APRC for the control-constrained system Σ.

When the decentralised MPC is implemented with the APRC-based attractivity constraint, the time responses are shown in Fig. 2.8, indicating attractive trajectories. The coefficient matrices Q_i, S_i, R_i of the quadratic constraint have been re-computed online using the first qualification condition (2.12), as detailed below.

The coefficient matrices Q_i, S_i, R_i of the supply rate can be updated online whenever $y_i^T Q_i y_i - \gamma_i \times \xi_i^- < 0$, as follows: The matrices $Q_i(k)$ and $R_i(k)$ will be determined from the feasible solutions to the following LMIs:

$$\begin{bmatrix} y_i^T[-Q_i + Q_i(k)]y_i & -y_i^T S_i \\ S_i^T y_i & -R_i + R_i(k) \end{bmatrix} \preccurlyeq 0, \quad R_i(k) \prec 0, \qquad (2.51a)$$

$$y_i^T Q_i(k)y_i - \gamma_i \times \xi_i^- > 0; \qquad (2.51b)$$

and choosing $S_i(k)$ as a null space of y_i^T, i.e. $S_i(k)y_i^T = 0$;

The LMI solver in MATLAB robust control toolbox has been used in this numerical example. The off-line coefficient matrices are as follows:

$$Q_1 = \begin{bmatrix} -11.7867 & -34.2257 & 0.3111 \\ -34.2257 & 21.5558 & -55.3959 \\ 0.3111 & -55.3959 & -72.5118 \end{bmatrix}, \quad Q_2 = \begin{bmatrix} -32.6274 & -11.8166 & -0.7111 \\ -11.8166 & -4.6037 & -3.8606 \\ -0.7111 & -3.8606 & -17.3838 \end{bmatrix},$$

$$S_1 = \begin{bmatrix} 5.7491 \\ -0.5833 \\ -13.5016 \end{bmatrix}, \quad S_2 = \begin{bmatrix} 5.9834 & 0.0064 & -3.3183 \\ 0.0064 & 22.5096 & -0.5098 \\ -3.3183 & -0.5098 & 24.4831 \end{bmatrix}, \quad S_3 = \begin{bmatrix} -4.5162 & 1.0599 \\ 1.0599 & -11.0225 \end{bmatrix},$$

$$R_1 = -77.0619, R_2 = -\begin{bmatrix} 75.6640 & 0.1853 & 0.5272 \\ 0.1853 & 80.2649 & -0.1551 \\ 0.5272 & -0.1551 & 83.0173 \end{bmatrix}, R_3 = -\begin{bmatrix} 76.3570 & -0.3167 \\ -0.3167 & 76.9403 \end{bmatrix}.$$

The chartering of a subsystem control shown in the above trajectories can be eliminated if we apply the off-line coefficient matrices when $\|x(k)\|$ lies inside a neighbourhood of zero, e.g. $\|x(k)\| \le 0.2$ in this numerical example.

2.5.3 Illustrative Example 3

The following subsystem state-space models are considered:

$$\dot{x}(t) = \begin{bmatrix} A_1 & & \\ & A_2 & \\ & & A_3 \end{bmatrix} x(t) + \begin{bmatrix} B_1 & & \\ & B_2 & \\ & & B_3 \end{bmatrix} u(t) + \begin{bmatrix} E_1 & & \\ & E_2 & \\ & & E_3 \end{bmatrix} v(t),$$

$$w(t) = \begin{bmatrix} F_1 & & \\ & F_2 & \\ & & F_3 \end{bmatrix} x(t), \quad v(t) = Hw(t), \quad y(t) = \begin{bmatrix} C_1 & & \\ & C_2 & \\ & & C_3 \end{bmatrix} u(t),$$

where

$$A_1 = \begin{bmatrix} -28.8 & 0 & 0 \\ 17.2 & -5.9 & 0 \\ -8.4 & 0 & 0 \end{bmatrix}, B_1 = \begin{bmatrix} 1 & 0 \\ 0 & 0 \\ 0 & 1 \end{bmatrix}, C_1 = \begin{bmatrix} 1 & 0 & 0 \\ 0 & 1 & 0 \\ 0 & 0 & 1 \end{bmatrix}, E_1 = \begin{bmatrix} 0 & -9.21 \\ 0 & 14.57 \\ 0 & 0 \end{bmatrix},$$

$$A_2 = \begin{bmatrix} -14.4 & 0 & 0 \\ 17.8 & -.6 & 0 \\ -3 & 0 & 0 \end{bmatrix}, B_2 = \begin{bmatrix} 0 \\ 0 \\ 1 \end{bmatrix}, C_2 = \begin{bmatrix} 1 & 0 & 0 \\ 0 & 1 & 0 \\ 0 & 0 & 1 \end{bmatrix}, E_2 = \begin{bmatrix} 18.44 & 0 & -12.35 & 0 \\ 0 & 0 & 0 & 0 \\ 0 & 1.47 & 0 & 0 \end{bmatrix},$$

$$A_3 = \begin{bmatrix} -16.3 & 0 & 0 \\ 11.9 & -.7 & 0 \\ -6.2 & 0 & 0 \end{bmatrix}, B_3 = \begin{bmatrix} -.1 & 0 \\ 1 & 0 \\ 0 & 1 \end{bmatrix}, C_3 = \begin{bmatrix} 1 & 0 & 0 \\ 0 & 1 & 0 \\ 0 & 0 & 1 \end{bmatrix}, E_3 = \begin{bmatrix} 23.64 & 0 \\ 0 & 0 \\ 0 & 1.81 \end{bmatrix},$$

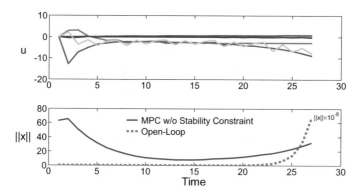

Fig. 2.9 Decentralised MPCs without APRC stability constraint

$$F_1 = \begin{bmatrix} 0.1 & 0 & 0 \\ 0 & 0 & 1 \end{bmatrix}, F_2 = \begin{bmatrix} 0.1 & 0 & 0 \\ 0 & 0 & 1 \end{bmatrix}, F_3 = \begin{bmatrix} 0 & 0.7 & 0 \end{bmatrix}, H = \begin{bmatrix} 0 & 0 & 1 & 0 & 0 \\ 0 & 0 & 0 & 1 & 0 \\ 1 & 0 & 0 & 0 & 1 \\ 0 & 1 & 0 & 0 & 0 \\ 0 & 0 & 1 & 0 & 0 \\ 0 & 0 & 0 & 1 & 0 \end{bmatrix}.$$

The weighting matrices in (2.29) are

$$\mathscr{R}_1 = \text{diag}\{2, 1\}, \ \mathscr{Q}_1 = \text{diag}\{5, 1, 2\}, \ \mathscr{R}_2 = 1,$$
$$\mathscr{Q}_2 = \text{diag}\{1, 1, 2\}, \ \mathscr{R}_3 = \text{diag}\{0.5, 1\}, \ \mathscr{Q}_3 = \text{diag}\{4, 1, 5\}.$$

The constraints on the control inputs are set up with

$$\|u_1\| \leqslant 2, \ \|u_2\| \leqslant 10, \ \|u_3\| \leqslant 10.$$

The initial states are as follows:

$$x_1(0) = \begin{bmatrix} 7 \\ 14 \\ -10 \end{bmatrix}, \ x_2(0) = \begin{bmatrix} -12 \\ -55 \\ 16 \end{bmatrix}, \ x_3(0) = \begin{bmatrix} -9 \\ 7 \\ -6 \end{bmatrix}.$$

Figure 2.9 depicts the time responses of the interconnected system regulated by the decentralised MPC. In this simulation study, the APRC has not been applied and the closed-loop system is unstable as a result.

When Procedure 2.2 is employed together with the online updated coefficient matrices as in (2.51) above, the time responses are shown in Fig. 2.10, indicating attractive trajectories. The Yalmip toolbox with the default SDP solver has been used in this simulation. The off-line coefficient matrices are as follows:

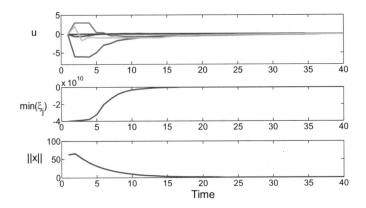

Fig. 2.10 Decentralised MPCs with APRCs and re-computed coefficient matrices

$$Q_1 = \begin{bmatrix} 4.1774 & 0.7683 & 1.3327 \\ 0.7683 & 0.5076 & -2.4361 \\ 1.3327 & -2.4361 & -0.8254 \end{bmatrix}, \quad Q_2 = \begin{bmatrix} 3.3861 & -1.1387 & 0.1554 \\ -1.1387 & -1.9627 & -0.2457 \\ 0.1554 & -0.2457 & -3.7694 \end{bmatrix},$$

$$S_1 = \begin{bmatrix} -0.1220 & 0.3023 \\ -0.1395 & 0.1755 \\ 0.2939 & -0.8126 \end{bmatrix}, \quad S_2 = \begin{bmatrix} 0.0295 & 0 \\ -0.1505 & 0 \\ -0.9662 & 0 \end{bmatrix}, \quad S_3 = \begin{bmatrix} -0.7497 & 0.2595 \\ -0.8920 & -0.1391 \\ -0.1594 & -0.9415 \end{bmatrix},$$

$$R_1 = -\begin{bmatrix} 3.4771 & 0.2019 \\ 0.2019 & 3.7955 \end{bmatrix}, \quad R_2 = -\begin{bmatrix} 3.6754 & 0 \\ 0 & 3.3593 \end{bmatrix}, \quad R_3 = -\begin{bmatrix} 3.8203 & 0.0483 \\ 0.0483 & 3.7358 \end{bmatrix}.$$

The time responses for a different set of initial states of the following:

$$x_1(0) = \begin{bmatrix} 1 \\ .7 \\ 3 \end{bmatrix}, \quad x_2(0) = \begin{bmatrix} .9 \\ .6 \\ 2 \end{bmatrix}, \quad x_3(0) = \begin{bmatrix} .2 \\ .8 \\ 4 \end{bmatrix}$$

are provided in Fig. 2.11. The system is also attractive, but the settling time is longer, compared to those in Fig. 2.10.

The centralised MPC using the *mpcmove* function from the MATLAB MPC toolbox is not feasible in this example due to the unstable system model. The eigenvalues of the state realisation matrix A are provided below.

$$0.0907, \ 0.7866, \ 0.9608,$$
$$2.2349, \ 1.2941, \ 0.4623,$$
$$-0.0004 + 0.0014i, \ -0.0004 - 0.0014i.$$

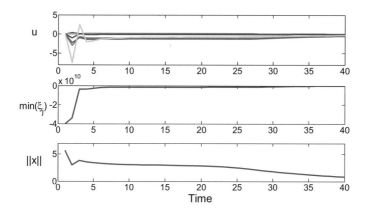

Fig. 2.11 Decentralised MPCs with APRCs and re-computed coefficient matrices—smaller initial states

2.6 Concluding Remarks

The asymptotic attractivity conditions for controlled interconnected systems in the discrete-time domain have been derived in this chapter based on the newly introduced quadratic constraints, including the asymptotically positive realness constraint (APRC), the quadratic dissipativity constraint (QDC) and the repeatedly energy-dissipative constraint (REDC).

The constructive approach of attractivity constraints for the decentralised MPC with those quadratic constraints has been delineated subsequently for practical implementations. The state constraints are not inclusive in the problem formulation herein, but only the control constraints are.

The parameters that will have to be tuned in this quadratic constraint method consist of the three coefficient matrices of the quadratic supply function, the rates of change of both the storage function and the supply rate, and the slack variable to avoid the optimising infeasibility.

Numerical simulations with a realistic helicopter model and two open-loop-unstable interconnected systems have been studied with MATLAB toolboxes. The results have demonstrated that the centralised and decentralised MPC schemes with the quadratic constraint (APRC or QDC) have guaranteed the controlled system asymptotic attractivity. Nevertheless, the asymptotic stability of the controlled global system around the zero equilibrium will usually be obtained from the centralised MPC with an adequately chosen predictive horizon when the corresponding open-loop system is stable.

Chapter 3
Quadratic Constraint for Parallel Splitting Systems

In this chapter, we consider parallel splitting connections in interconnected systems having mixed connection structures, as shown in Fig. 3.1. In these subsystems, the coupling vectors v_i and w_i are split into two subcoupling vectors to explicitly describe the serial and parallel splitting connections. Unlike the parallel systems described in the literature of dissipative system theory wherein each input also equals to the common input, see, e.g. [139], here the parallel splitting, or parallelised, systems whose inputs are summed up to a common input, as shown in Fig. 3.2, are considered.

For a plant-wide system that has multiple concurrent processing streams with subprocesses, this parallelised connection is suitable for modelling those subprocesses independently together with a modified global coupling matrix (adjacent matrix). As a result, each subprocess can be installed with a local MPC optimisation within a decentralised MPC scheme for the plant-wide global system.

3.1 System and Control Model

Consider an interconnected system Σ consisting of h units, denoted as $\mathcal{G}_j$, $j = 1, 2, \ldots, h$, as depicted in Fig. 3.1. Each unit $\mathcal{G}_j$ is represented by

$$\mathcal{G}_j : \begin{cases} x_j(k+1) = A_j\, x_j(k) + B_j\, u_j(k) + E_{\Theta_j}\, v_{\Theta_j}(k) + E_{\Delta_j}\, v_{\Delta_j}(k), \\ \quad\ y_j(k) = C_j\, x_j(k), \\ \quad\ w_{\Theta_j}(k) = F_{\Theta_j}\, x_j(k), \\ \quad\ w_{\Delta_j}(k) = F_{\Delta_j}\, x_j(k) \end{cases} \tag{3.1}$$

where x_j, u_j and y_j are the state, control input and measurement output of each unit, respectively.

© Springer Nature Singapore Pte Ltd. 2018
A. Tri Tran C. and Q. Ha, *A Quadratic Constraint Approach to Model Predictive Control of Interconnected Systems*, Studies in Systems, Decision and Control 148, https://doi.org/10.1007/978-981-10-8409-6_3

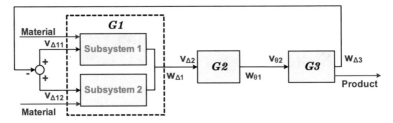

Fig. 3.1 Mixed connection structure with parallelised subsystems of an interconnected system

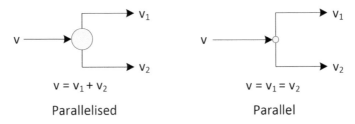

Fig. 3.2 Parallel splitting (or parallelised) and parallel connections

There are two types of interactive inputs and outputs, parallelised and serial, in this model. The serial interactive input and output are v_{Θ_j} and w_{Θ_j}, while the parallelised interactive input and output are v_{Δ_j} and w_{Δ_j}, respectively.

3.1.1 Serial Connection

Unit $\mathscr{G}_i$ is said to be *connected serially* to unit $\mathscr{G}_j$ if there is a nonzero coupling matrix H_{Θ}^{ij} with zero or one entries such that $v_{\Theta j}(k) = H_{\Theta}^{ij} w_{\Theta i}$. If $\mathscr{G}_i$ is not serially connected to $\mathscr{G}_j$, set $H_{\Theta}^{ij} = 0$. Accordingly,

$$H_{\Theta} = [H_{\Theta}^{ij}]_h$$

is regarded as the global serial coupling matrix for $v_{\Theta}(k) = H_{\Theta} w_{\Theta}(k)$, where $v_{\Theta}(k) := [v_{\Theta 1}^T(k) \ldots v_{\Theta h}^T(k)]^T$, $w_{\Theta}(k) := [w_{\Theta 1}^T(k) \ldots w_{\Theta h}^T(k)]^T$. For instance, in Fig. 3.1 unit $\mathscr{G}_2$ is connected in serial to unit $\mathscr{G}_3$.

3.1.2 Parallelised Connection

Unit $\mathscr{G}_i$ is said to be *connected 'parallelisedly'* to unit $\mathscr{G}_j$ if there is a nonzero coupling matrix H_{Δ}^{ij} of zero or one entries such that $w_{\Delta i}(k) = H_{\Delta}^{ij} v_{\Delta_j}$. If $\mathscr{G}_i$ is not parallelisedly connected to $\mathcal{G}_j$, simply set $H_\Delta^{ij} = 0$. Accordingly,

$$H_\Delta = [H_\Delta^{ij}]_h$$

is regarded as the global parallelised coupling matrix for $w_\Delta(k) = H_\Delta v_\Delta$, where $v_\Delta(k) := [v_{\Delta 1}^T(k) \dots v_{\Delta h}^T(k)]^T$, $w_\Delta(k) := [w_{\Delta 1}^T(k) \dots w_{\Delta h}^T(k)]^T$. For instance, in Fig. 3.1 unit $\mathcal{G}_3$ is connected to unit $\mathcal{G}_1$ via the parallelised input and output:

$$w_{\Delta 3}(k) = v_{\Delta 11}(k) + v_{\Delta 12}(k) = [1 \quad 1]v_{\Delta 1}(k)$$

for $v_{\Delta 1}(k) = [v_{\Delta 11}^T(k) \ v_{\Delta 12}^T(k)]^T$.

The splitting ratios of $w_{\Delta 3}(k)$ to $v_{\Delta 11}(k)$ and $v_{\Delta 12}(k)$ are unknown in this work.

3.1.3 Global System

With the definition of

$$A := \mathsf{diag}\,[A_j]_1^h, \ B := \mathsf{diag}\,[B_j]_1^h, \ E_\Theta := \mathsf{diag}\,[E_{\Theta_j}]_1^h,$$
$$E_\Delta := \mathsf{diag}\,[E_{\Delta_j}]_1^h, \ C := \mathsf{diag}\,[C_j]_1^h, \ F_\Theta := \mathsf{diag}\,[F_{\Theta_j}]_1^h, \ F_\Delta := \mathsf{diag}\,[F_{\Delta_j}]_1^h,$$
$$x(k) := [x_1^T(k) \dots x_h^T(k)]^T, \ u(k) := [u_1^T(k) \dots u_h^T(k)]^T,$$

the larger scale global system has the state-space model of

$$\Sigma_\Delta : \begin{cases} x(k+1) = (A + E_\Theta H_\Theta F_\Theta)x(k) + Bu(k) + E_\Delta v_\Delta(k), \\ \quad y(k) = C\,x(k), \\ \quad w_\Delta(k) = F_\Delta\,x(k), \\ \quad w_\Delta(k) = H_\Delta v_\Delta(k). \end{cases} \tag{3.2}$$

The same control constraint (2.2) is considered for $u_j(k)$; i.e., each unit $\mathcal{G}_j$ is installed with a local MPC optimisation in a decentralised MPC scheme. There may be some subsystems within a unit, such as $\mathcal{G}_1$ in Fig. 3.1, but the local MPC optimisation is not for each individual subsystem, but for the unit as a whole.

3.2 Parallel Splitting System with a Matrix Annihilation

Denote the null base of $\begin{bmatrix} F_\Delta & -H_\Delta \end{bmatrix}$ as $\mathcal{N}$. If Q_Δ, S_Δ, R_Δ are such that

$$\mathcal{N}^T \begin{bmatrix} Q_\Delta & S_\Delta \\ S_\Delta^T & R_\Delta \end{bmatrix} \mathcal{N} \geqslant 0, \tag{3.3}$$

then

$$\xi_\Delta(k) := \begin{bmatrix} x(k) \\ v_\Delta(k) \end{bmatrix}^T \begin{bmatrix} Q_\Delta & S_\Delta \\ S_\Delta^T & R_\Delta \end{bmatrix} \begin{bmatrix} x(k) \\ v_\Delta(k) \end{bmatrix} \geqslant 0 \tag{3.4}$$

for all $(x(k), v_\Delta(k))$ satisfying $H_\Delta v_\Delta(k) = F_\Delta x(k)$.

Define the quadratic function of

$$\xi_j(y_j, u_j) := \begin{bmatrix} y_j \\ u_j \end{bmatrix}^T \begin{bmatrix} X_j & Y_j \\ Y_j^T & Z_j \end{bmatrix} \begin{bmatrix} y_j \\ v_j \end{bmatrix} \tag{3.5}$$

for symmetric matrices X_j and Z_j and matrix Y_j of appropriate dimension. Accordingly,

$$\xi(y, u) := \sum_{j=1}^h \xi_j = \begin{bmatrix} y \\ u \end{bmatrix}^T \begin{bmatrix} X & Y \\ Y^T & Z \end{bmatrix} \begin{bmatrix} y \\ u \end{bmatrix},$$

with $X := \mathsf{diag}[X_j]_1^h$, $Y := \mathsf{diag}[Y_j]_1^h$, $Z := \mathsf{diag}[Z_j]_1^h$.

An enhanced version of the asymptotically positive realness constraint is introduced in this chapter.

3.2.1 Asymptotically Surely Positive Realness Constraint and Attractability Condition

The asymptotically surely positive realness constraint (ASPRC) is defined as follows:

Definition 3.1 The controlled motion $(y_j(k), u_j(k))$ of a unit $\mathcal{G}_j$ (3.1) is said to be *asymptotically surely positive real* with respect to $\xi_j(k) := \xi_j(y_j(k), u_j(k))$, or simply ASPR, if there exists an initial time k_0, such that $\xi_j(k_0) < 0$ and

$$\xi_j(k) \to o_j \geqslant 0 \text{ as } k \to +\infty, \tag{3.6}$$

and $o_j \to 0$ as $k \to +\infty$.

One can see that inequality (3.6) is an enhanced version of the APRC (2.8).

Lemma 3.1 The controlled motion $(y_j(k), u_j(k))$ of a unit $\mathcal{G}_j$ is ASPR with respect to $\xi_j(k)$, if there are $k_0 \in \mathbb{Z}^+$, $0 \leqslant \gamma_j < 1$, and $\varepsilon_j(k) \geqslant 0$, such that the following asymptotically surely positive realness constraint (ASPRC) holds true for all $k > k_0$:

$$ASPRC: \begin{cases} \xi_j(k) \geqslant \gamma_j \xi_j(k-1) + \varepsilon_j(k), \\ 0 > \xi_j(k_0), \end{cases} \tag{3.7}$$

for $\varepsilon_j(k) \to 0$ as $k \to +\infty$.

Proof The proof is similar to that for Lemma 2.1 with $o_j = \varepsilon_j \dfrac{1-\gamma_j^{k-k_0}}{1-\gamma_j}$. ∎

The asymptotic attractability condition for Σ (3.2) is stated below.
Define $A_\Theta := A + E_\Theta H_\Theta F_\Theta$ and $\mathcal{M}(P, Q_\Delta, S_\Delta, R_\Delta, X, Y, Z) :=$

$$
\begin{bmatrix}
A_\Theta^T P A_\Theta - \tau P + Q_\Delta + C^T X C & * & * \\
B^T P A_\Theta + Y^T C & B^T P B + Z & * \\
E_\Delta^T P A_\Theta + S_\Delta^T F_\Delta & E_\Delta^T P B & E_\Delta^T P E_\Delta + R_\Delta
\end{bmatrix}. \tag{3.8}
$$

Theorem 3.1 *Let $0 < \tau < 1$. Suppose that for P, X, Y, Z solving the following LMIs:*

$$
\mathcal{M}(P, Q_\Delta, S_\Delta, R_\Delta, X, Y, Z) \prec 0, \quad P \succ 0, \tag{3.9}
$$

the decoupled control sequences $\{u_j \in \mathbb{R}^{m_j}\}$, $j = 1, 2, \ldots, h$, are bounded by the respective ASPRCs (3.7), and there exists a compact neighbourhood of the zero equilibrium $\mathbb{X}$ such that Assumption 1 holds for Σ_Δ (3.2) in $\mathbb{X}$ with some of h decoupled control sequences $\{u_j \in \mathbb{R}^{m_j}\}$ among the above ASPRC-bounded control sequences.

Then, the unconstrained controlled system Σ_Δ (3.2) is asymptotically attractive.

Proof Let $V(x) := x^T P x$, $P \succ 0$. By (3.9),

$$
\begin{aligned}
V(x(k+1)) - \tau V(x(k)) + \xi_\Delta(k) + \xi(k) &= \\
\begin{bmatrix} x(k) \\ u(k) \\ v_\Delta(k) \end{bmatrix}^T \mathcal{M}(P, Q_\Delta, S_\Delta, R_\Delta, X, Y, Z) \begin{bmatrix} x(k) \\ u(k) \\ v_\Delta(k) \end{bmatrix} &\leqslant 0.
\end{aligned} \tag{3.10}
$$

As $\xi_\Delta(k) \geqslant 0 \; \forall k$ using the annihilation (3.3), and (3.7), we obtain

$$
V(x(k+1)) \leqslant \tau V(x(k)) - \gamma \xi(k-1) - o, \quad o = \sum_{j=1}^h o_j \geqslant 0. \tag{3.11}
$$

Similar to the proof for Theorem 2.1, we conclude that $\|x(k)\| \to 0$ as $k \to +\infty$. ∎

It is clear that, Theorem 3.1 remains valid if $\xi_j(u_j, y_j)$ in the ASPRC (3.7) is replaced by $\delta_j(u_j, y_j)$,

$$
\delta_j(u_j, y_j) := \begin{bmatrix} y_j \\ u_j \end{bmatrix}^T \begin{bmatrix} Q_j & S_j \\ S_j^T & R_j \end{bmatrix} \begin{bmatrix} y_j \\ u_j \end{bmatrix},
$$

and

$$
\xi_j(y_j, u_j) = \begin{bmatrix} y_j \\ u_j \end{bmatrix}^T \begin{bmatrix} X_j & Y_j \\ Y_j^T & Z_j \end{bmatrix} \begin{bmatrix} y_j \\ u_j \end{bmatrix} \geqslant \delta_j(u_j, y_j) = \begin{bmatrix} y_j \\ u_j \end{bmatrix}^T \begin{bmatrix} Q_j & S_j \\ S_j^T & R_j \end{bmatrix} \begin{bmatrix} y_j \\ u_j \end{bmatrix} \quad \forall u_j,
$$

which is equivalent to an LMI in Q_j, S_j, R_j of

$$\begin{bmatrix} y_j^T(X_j - Q_j)y_j & y_j^T(Y_j - S_j) \\ (Y_j - S_j)^T y_j & Z_j - R_j \end{bmatrix} \succcurlyeq 0. \tag{3.12}$$

The following constraint is now considered:

$$\delta_j(\boldsymbol{u}_j, y_j) \geqslant \gamma_j \xi_j^- + \varepsilon_j \; \forall k > k_0, \tag{3.13}$$

instead of (3.7).

For the practical implementations, select a small tuning number $\mu_j > 0$. Then,

1. Whenever $\xi_j^- < -\mu_j$ set

$$\varphi_j^- := \xi_j^-, \; S_j^T \equiv \mathcal{N}_{y_j^T}, \tag{3.14}$$

where $\mathcal{N}_{y_j^T}$ stands for the basis of the null space of y_j^T. Then (3.12) reads

$$\begin{bmatrix} y_j^T(X_j - Q_j)y_j & y_j^T Y_j \\ Y_j^T y_j & Z_j - R_j \end{bmatrix} \succcurlyeq 0 \tag{3.15}$$

and (3.13) is the following quadratic constraint in $\boldsymbol{u}_j(k)$,

$$\boldsymbol{u}_j^T R_j \boldsymbol{u}_j \geqslant -y_j^T Q_j y_j + \gamma_j \xi_j^- + \varepsilon_j. \tag{3.16}$$

Obviously, the condition

$$y_j^T Q_j y_j - \gamma_j \xi_j^- - \varepsilon_j \geqslant 0 \tag{3.17}$$

is needed for a solution to exist for (3.16);

2. Whenever $\xi_j^- \geqslant -\mu_j$, (3.13) is replaced by the following convex constraint in $\boldsymbol{u}_j$

$$\xi_j := \begin{bmatrix} y_j \\ \boldsymbol{u}_j \end{bmatrix}^T \begin{bmatrix} X_j & Y_j \\ Y_j^T & Z_j \end{bmatrix} \begin{bmatrix} y_j \\ \boldsymbol{u}_j \end{bmatrix} \geqslant 0. \tag{3.18}$$

3. Whenever $\xi_j^- \geqslant 0$, the system is deemed attractive.

The asymptotic attractivity condition is stated as follows:

Proposition 3.1 *Let $\mu_j > 0$. Suppose that, for P, Q_Δ, R_Δ, S_Δ, X, Z, and Y solving LMIs (3.9) and (3.3), and the matrices R_j, Q_j and $S_j^T \equiv \mathcal{N}_{y_j^T}$ subsequently solving (3.15) and (3.17) whenever $\xi_j^- < -\mu_j$, the decoupled controls $\{u_j \in \mathbb{U}_j\}$, $j = 1, 2, \ldots, h$, are bounded by the respective conditional convex constraints of*

$$\begin{cases} (3.16), \textit{ for } \xi_j^- < -\mu_j, \\ (3.18), \textit{ for } \xi_j^- \geqslant -\mu_j, \end{cases} \qquad (3.19)$$

and there exists a compact neighbourhood of the zero equilibrium $\mathbb{X}$ *such that Assumption 1 holds for* Σ_Δ *(3.2) in* $\mathbb{X}$ *with some h decoupled control sequences* $\{u_j \in \mathbb{U}_j\},\ j = 1, 2, \ldots, h$ *among the above ASPRC–bounded control sequences; Then the constrained controlled system* Σ_Δ *(3.2) is asymptotically attractive.*

Proof This is a direct result of Theorem 3.1 and the above analysis. ■

The decentralised MPC problem with the ASPRC for interconnected systems having serial and parallelised connections is addressed in the next section.

3.2.2 Decentralised MPC for Parallel Splitting Systems

With a predictive horizon N, a local MPC optimisation with the ASPRC is as follows:

$$\min_{\hat{u}_j(k+\ell)} \sum_{\ell=1}^{N+1} x_j^T(k+\ell)\mathcal{Q}_j x_j(k+\ell) + \sum_{\ell=0}^{N} u_j^T(k+\ell)\mathcal{R}_j u_j(k+\ell) \quad (3.20)$$

subject to $x_j(k+\ell+1) = A_j x_j(k+\ell) + B_j u_j(k+\ell),\ \ell = 0, 1, \ldots, N,$

$$(3.16) \text{ for } j \in I^-(k) \text{ and } (3.18) \text{ for } j \in I^+(k)$$

where
$I^-(k) := \{j \in \{1, 2, \ldots, h\} : \xi_j^- < -\mu_j\},\ I^+ := \{1, 2, \ldots, h\} \setminus I^-(k) \bigcap I^{++}(k)$, in which $I^{++}(k) := \{j \in \{1, 2, \ldots, h\} : y_j^T(X_j - Y_j Z_j^{-1} Y_j^T)y_j \geqslant 0\}.$

For each unit, the optimisation problem (3.20) is solved by the local optimiser for the minimising sequence $\hat{u}_j^*(k+\ell),\ j = 0, 1, \ldots, N$, but only its first element $u_j^*(k)$ is applied to control the plant. This rolling process is repeated at the next time step and continues thereon.

Another method of resolving the parallelised connections is outlined in the next section. The matrix annihilation is not employed in this so-called parallelised masking method.

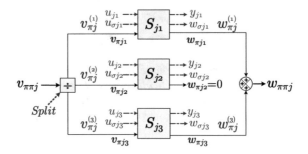

Fig. 3.3 Parallel splitting of a unit $\mathscr{G}_j$ having three subsystems

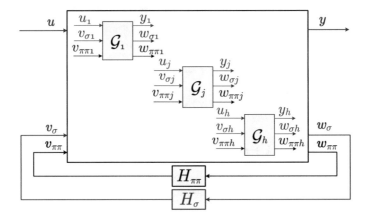

Fig. 3.4 Block diagram of an interconnected system Σ_Δ on the basis of units $\mathscr{G}_j$

3.3   Parallelised Masking Dissipativity Criterion

A fully decentralised dissipativity criterion for interconnected systems having parallel splitting and serial connections is given in this section. A hierarchical view of such parallel splitting and serial connections is shown in Figs. 3.3 and 3.4.

3.3.1   Subsystem Control Model

Consider an interconnected system Σ_Δ consisting of (i) h units, denoted as $\mathscr{G}_j$, $j = 1, 2, \ldots, h$; (ii) each unit $\mathscr{G}_j$ has g_j subsystems, denoted as $\mathscr{S}_i$, $i = 1, 2, \ldots, g_j$.

A subsystem $\mathscr{S}_i$ is represented by the discrete-time state-space model of the form

$$\mathscr{S}_i : \begin{cases} x_i(k+1) = A_i\, x_i(k) + B_i\, u_i(k) + E_i\, v_i(k), \\ y_i(k) = C_i\, x_i(k), \quad w_i(k) = F_i\, x_i(k), \end{cases} \tag{3.21}$$

$$E_i = [E_{\pi ji}\ E_{\sigma ji}],\ F_i = [F_{\pi ji}\ F_{\sigma ji}],\ v_i = [v_{\pi ji}^T\ v_{\sigma ji}^T]^T,\ w_i = [w_{\pi ji}^T\ w_{\sigma ji}^T]^T,$$

where

- $v_{\sigma ji} \in \mathbb{V}_i \subset \mathbb{R}^{m_{v_{\sigma ji}}}$ and $w_{\sigma ji} \in \mathbb{W}_i \subset \mathbb{R}^{q_{w_{\sigma ji}}}$ are *serial coupling* input and output vectors, respectively,
- $v_{\pi ji} \in \mathbb{R}^{m_{v_{\pi j}}}$ and $w_{\pi ji} \in \mathbb{R}^{q_{w_{\pi j}}}$ are *parallelised coupling* input and output vectors, respectively,
- $x_i \in \mathbb{X}_i \subset \mathbb{R}^{n_i}$ is the state vector, $y_i \in \mathbb{Y}_i \subset \mathbb{R}^{q_i}$ is the measurement output vector, and $u_i \in \mathbb{U}_i \subset \mathbb{R}^{m_i}$ is the control input vector.

There are $h_{\Sigma_\Delta} := \sum_{j=1}^h g_j$ subsystems $\mathscr{S}_i$ in Σ_Δ. The subsystem control constraint is also considered in this section, i.e.

$$\mathbb{U}_i := \{u_i : \|u_i\|^2 \leqslant \rho_i\}. \tag{3.22}$$

Serial connections: Two subsystems $\mathscr{S}_{\xi i}$ and $\mathscr{S}_{\zeta o}$ are said to be *serially connected* (SC) if the coupling input vector $v_{\sigma \xi i}$ of $\mathscr{S}_{\xi i}$ is the coupling output vector $w_{\sigma \zeta o}$ of $\mathscr{S}_{\zeta o}$, i.e.

$$\text{(SC)}\quad v_{\sigma \xi i}(k) = w_{\sigma \zeta o}(k). \tag{3.23}$$

3.3.2 Unit Control Model

In the sequel, M_j denotes the diagonal matrix $\text{diag}[M_i]_1^{g_j}$, and v_{*j} denotes the stacking vector $[v_{*j1}^T \ldots v_{*jg_j}^T]^T$ (i.e. only the subscript j denoting a unit retains). A unit $\mathscr{G}_j$ is represented by the block-diagonal system formed by g_j parallelised subsystems $\mathscr{S}_i$ as

$$\mathscr{G}_j : \begin{cases} x_j(k+1) = A_j x_j(k) + B_j u_j(k) + E_j v_j(k), \\ y_j(k) = C_j x_j(k), \quad w_j(k) = F_j x_j(k), \end{cases} \tag{3.24}$$

$$E_j = [E_{\pi j}\ E_{\sigma j}],\ F_j = [F_{\pi j}\ F_{\sigma j}],\ v_j^T = [v_{\pi j}^T\ v_{\sigma j}^T],\ w_j^T = [w_{\pi j}^T\ w_{\sigma j}^T].$$

where

$$A_j = \text{diag}[A_i]_1^{g_j},\ B_j = \text{diag}[B_i]_1^{g_j},\ C_j = \text{diag}[C_i]_1^{g_j},\ E_{\pi j} = \text{diag}[E_{\pi ji}]_1^{g_j},$$
$$F_{\pi j} = \text{diag}[F_{\pi ji}]_1^{g_j},\ E_{\sigma j} = \text{diag}[E_{\sigma ji}]_1^{g_j},\ F_{\sigma j} = \text{diag}[F_{\sigma ji}]_1^{g_j},$$
$$x_j^T = [x_{j1}^T \ldots x_{jg_j}^T],\ u_j^T = [u_{j1}^T \ldots u_{jg_j}^T],\ y_j^T = [y_{j1}^T \ldots y_{jg_j}^T],\ v_{\sigma j}^T = [v_{\sigma j1}^T \ldots v_{\sigma jg_j}^T],$$
$$w_{\sigma j}^T = [w_{\sigma j1}^T \ldots w_{\sigma jg_j}^T],\ v_{\pi j}^T = [v_{\pi j1}^T \ldots v_{\pi jg_j}^T],\ w_{\pi j}^T = [w_{\pi j1}^T \ldots w_{\pi jg_j}^T].$$

Parallelised connections within a unit: The parallelised coupling vectors $v_{\pi ji}$ and $w_{\pi ji}$ of all subsystems $\mathscr{S}_i$, $i = 1, 2, \ldots, g_j$, belonging to a unit $\mathscr{G}_j$, $j \in \{1, 2, \ldots, h\}$, are assumed, without loss of generality, having the same size. If there is only one parallelised signal, the block diagram of the parallelised connections within a unit $\mathcal{G}_j$ having three subsystems ($g_j = 3$) is given in Fig. 3.3. The divider operator at $v_{\pi\pi j}$ in this figure represents the splitting of $v_{\pi\pi j}$ into $v_{\pi j}^{(\ell)}$.

Two new signals, $v_{\pi\pi j}$ and $w_{\pi\pi j}$, are introduced here, while the outputs $w_{\pi j}$ are summed up to become $w_{\pi\pi j}$ and the input $v_{\pi\pi j}$ is split up into $v_{\pi j}$. Their relationships are represented by two matrices Ψ_{v_j} and Ψ_{w_j}, which are defined as

$$\text{(PC)}\begin{array}{l} v_{\pi\pi j} := v_{\pi j1} + \cdots + v_{\pi jg_j} := \Psi_{v_j} v_{\pi j}, \\ w_{\pi\pi j} := w_{\pi j1} + \cdots + w_{\pi jg_j} := \Psi_{w_j} w_{\pi j}, \end{array} \qquad (3.25)$$

where $v_{\pi j}^T = [v_{\pi j1}^T \ldots v_{\pi jg_j}^T]$, $w_{\pi j}^T = [w_{\pi j1}^T \ldots w_{\pi jg_j}^T]$.

For the case of parallel-redundant subprocesses whose material or energy flows are split into smaller flows to each subprocess which is a subsystem $\mathcal{S}_i$ here, it is impossible to have a constant splitting ratio at all time, but dynamic ratios. The splitting ratios between $v_{\pi ji}$ of $v_{\pi j}$ should, therefore, be *unknown*. Due to these unknown splitting ratios of parallel-redundant subsystems with parallelised connections, the connectivity of the larger scale interconnection process cannot be established on the basis of subsystems. In later subsections, we tackle this issue by establishing the dissipativity criteria for a unit consisting of parallelised subsystems; i.e., the open-loop subsystems are not necessarily dissipative, but only their auspice unit is. It is noted here that the common input and output vectors, $v_{\pi\pi j}$ and $w_{\pi\pi j}$, of parallelised connections are involved in the global connection matrix, instead of $v_{\pi j}$ and $w_{\pi j}$.

3.3.3 Global System Control Model

In what follows, M (of Σ_Δ) denotes the diagonal matrix $\text{diag}[M_j]_1^h$, and v_* denotes the stacking vector $[v_{*1}^T \ldots v_{*h}^T]^T$ (unit subscripts j vanish). The interconnected system Σ_Δ is represented by the block-diagonal system formed by h diagonal units $\mathcal{G}_j$ (or h_{Σ_Δ} subsystems $\mathcal{S}_i$), and the interconnection processes H_σ, $H_{\pi\pi}$ as

$$\Sigma_\Delta : \begin{cases} x(k+1) = A\,x(k) + B\,u(k) + Ev(k) \\ y(k) = Cx(k), \; w(k) = Fx(k), \\ v_{\pi\pi}(k) = H_{\pi\pi}\, w_{\pi\pi}(k), \; v_\sigma(k) = H_\sigma\, w_\sigma(k). \end{cases} \qquad (3.26)$$

where

$$E = [E_\pi \; E_\sigma], \; F = [F_\pi \; F_\sigma], \; v^T = [v_\pi^T \; v_\sigma^T], \; w^T = [w_\pi^T \; w_\sigma^T].$$

Specifically,

$$A = \text{diag}[A_j]_1^h, \ B = \text{diag}[B_j]_1^h, \ C = \text{diag}[C_j]_1^h, \ E_\pi = \text{diag}[E_{\pi j}]_1^h,$$
$$F_\pi = \text{diag}[F_{\pi j}]_1^h, \ E_\sigma = \text{diag}[E_{\sigma j}]_1^h, \ F_\sigma = \text{diag}[F_{\sigma j}]_1^h, \ x^T = [x_1{}^T \ldots x_h^T],$$
$$u^T = [u_1{}^T \ldots u_h^T], \ y^T = [y_1^T \ldots y_h^T], \ v_\pi^T = [v_{\pi 1}^T \ldots v_{\pi h}^T], \ w_\pi^T = [w_{\pi 1}^T \ldots w_{\pi h}^T],$$
$$v_{\pi\pi}^T = [v_{\pi\pi 1}^T \ldots v_{\pi\pi h}^T], \ w_{\pi\pi}^T = [w_{\pi\pi 1}^T \ldots w_{\pi\pi h}^T], \ v_\sigma^T = [v_{\sigma 1}^T \ldots v_{\sigma h}^T],$$
$$w_\sigma^T = [w_{\sigma 1}^T \ldots w_{\sigma h}^T].$$

The parallelised connection process inside Σ_Δ is as follows:

$$v_{\pi\pi}(k) = \Psi_v \, v_\pi(k), \ w_{\pi\pi}(k) = \Psi_w \, w_\pi(k), \tag{3.27}$$

where $\Psi_v := \text{diag}[\Psi_{v_j}]_1^h, \ \Psi_w := \text{diag}[\Psi_{w_j}]_1^h$.

The interconnections between units and subsystems are specified by the interconnection matrices $H_{\pi\pi}$ and H_σ, respectively. The elements of $H_{\pi\pi}$ and H_σ are zero or one only. The block diagrams of the larger scale global system Σ_Δ from the unit perspective is depicted in Fig. 3.4.

3.3.4 Subsystem Stand-Alone Control Model

In a decentralised architecture, the stand-alone subsystem model (when all coupling inputs and outputs vanish: $v_i = 0, w_i = 0$) is also requisite and represented by

$$\mathscr{S}_i|_{\text{stand alone}} : \begin{cases} x_i(k+1) = A_i \, x_i(k) + B_i \, u_i(k), \\ y_i(k) = C_i \, x_i(k). \end{cases} \tag{3.28}$$

Assumptions: (i) It is assumed that the fast changing-over between the duty and standby subsystems within one unit is smooth; i.e., sudden state jumps will not incur frequently due to the changing-over operations. We assume here that the occasional state jumps will not, however, cause the system instability. The controls for duty and standby subsystems are regarded as internal activities of that unit. (ii) As mentioned in the first chapter, the issue of unknown splitting ratios in parallelised connections is tackled by having a combined solution, wherein the APRC is developed for each individual subsystem $\mathscr{S}_i$ while the dissipativity criterion is derived for the unit $\mathscr{G}_j$ consisting of g_j parallelised subsystems $\mathscr{S}_i$.

3.3.5 Dissipative and Attractive Conditions

For a unit $\mathscr{G}_j$, consider the quadratic supply rate $\xi_j(y_{jD}, u_{jD})$ defined as

$$
\xi_j\big(y_{jD}(k), u_{jD}(k)\big) := \begin{bmatrix} y_{jD}(k) \\ u_{jD}(k) \end{bmatrix}^T \begin{bmatrix} Q_j & S_j \\ S_j^T & R_j \end{bmatrix} \begin{bmatrix} y_{jD}(k) \\ u_{jD}(k) \end{bmatrix}, \tag{3.29}
$$

where $y_{jD}^T = [y_j^T w_{\pi\pi j}^T w_{\sigma j}^T]$, $u_{jD}^T = [u_j^T v_{\pi\pi j}^T v_{\sigma j}^T]$, $Q_j = \mathrm{diag}\{Q_{11j}, Q_{22j}, Q_{33j}\}$,

$$
S_j = \mathrm{diag}\{S_{11j}, S_{22j}, S_{33j}\}, \quad R_j = \mathrm{diag}\{R_{11j}, R_{22j}, R_{33j}\},
$$

in which each block element has a block-diagonal structure corresponding to subsystem dimensions with symmetric Q_j and R_j.

Definition 3.2 A unit $\mathscr{G}_j$ is said to be quadratically dissipative with respect to $\xi_j(y_{jD}, u_{jD})$ if there exists a nonnegative C^1 function, addressed as storage function, $V_j(x_j(k))$, $V_j(0) = 0$, such that for all u_{jD} and all $k \in \mathbb{Z}^+$, the following dissipation inequality is satisfied irrespectively of the initial value of the state $x_j(0)$:

$$
V_j\big(x_j(k+1)\big) - \tau_j\, V_j\big(x_j(k)\big) \leqslant \xi_j\big(y_{jD}(k), u_{jD}(k)\big), \; 0 < \tau_j < 1. \tag{3.30}
$$

The storage function $V_j\big(x(k)\big) = x_j(k)^T P_j x_j(k)$, $P_j = P_j^T \succ 0$, is considered in this section.

Similar to Chap. 2, the decoupled APRC-based stability constraint in u_i for each subsystem $\mathscr{S}_i$ of the form

$$
u_i^T Q_{ic} u_i + 2y_i^T S_{ic}^T u_i \geqslant \gamma_i\, \xi_i^- - y_i^T R_{ic} y_i, \; 0 < \gamma_i < 1, \tag{3.31}
$$

where $Q_{ic} \equiv -R_{11i}$, $S_{ic} \equiv -S_{11i}$, $R_{ic} \equiv -Q_{11i}$, is considered in this section.

This means each subsystem $\mathscr{S}_i$ is dissipative with respect to $\xi_i(y_i(k), u_i(k))$, $i = 1, 2, \ldots, h_{\Sigma_\triangle}$.

Define the dissipativity matrices of the larger scale global system $\Sigma_\triangle$ as

$$
Q_{11} = \mathrm{diag}[Q_{11j}]_1^h, \; S_{11} = \mathrm{diag}[S_{11j}]_1^h, \; R_{11} = \mathrm{diag}[R_{11j}]_1^h,
$$

$$
Q_{22} = \mathrm{diag}[Q_{22j}]_1^h, \; S_{22} = \mathrm{diag}[S_{22j}]_1^h, \; R_{22} = \mathrm{diag}[R_{22j}]_1^h,
$$

$$
Q_{33} = \mathrm{diag}[Q_{33j}]_1^h, \; S_{33} = \mathrm{diag}[S_{33j}]_1^h, \; R_{33} = \mathrm{diag}[R_{33j}]_1^h.
$$

The dissipativity matrices of unit $\mathscr{G}_j$ are

$$
Q_{11j} = \mathrm{diag}[Q_{11i}]_1^{g_j}, \; Q_{22j} = \mathrm{diag}[Q_{22i}]_1^{g_j}, \; Q_{33j} = \mathrm{diag}[Q_{33i}]_1^{g_j},
$$

$$
S_{11j} = \mathrm{diag}[S_{11i}]_1^{g_j}, \; S_{22j} = \mathrm{diag}[S_{22i}]_1^{g_j}, \; S_{33j} = \mathrm{diag}[S_{33i}]_1^{g_j},
$$

$$R_{11j} = \text{diag}[R_{11i}]_1^{g_j}, \quad R_{22j} := \text{diag}[R_{22i}]_1^{g_j}, \quad R_{33j} := \text{diag}[R_{33i}]_1^{g_j}.$$

where Q_{11i}, R_{11i} are symmetric matrices of dimensions q_i, m_i, respectively; Q_{22i}, R_{22i} are symmetric matrices of dimensions $q_{w_{\pi j}}$, $m_{v_{\pi j}}$, respectively; Q_{33i}, R_{33i} are symmetric matrices of dimensions $q_{w_{\sigma ji}}$, $m_{v_{\sigma ji}}$, respectively; S_{11i} are rectangular matrices of $q_i \times m_i$ dimensions; S_{22i} are rectangular matrices of $q_{w_{\pi j}} \times m_{v_{\pi j}}$ dimensions; S_{33i} are rectangular matrices of $q_{w_{\sigma ji}} \times m_{v_{\sigma ji}}$ dimensions.

The asymptotic attractivity condition for the larger scale global system based on the QDC is stated next.

Theorem 3.2 *Let* $0 < \tau_j < 1$. *For* P_j, Q_{11i}, S_{11i}, R_{11i}, Q_{22j}, S_{22j}, R_{22j}, Q_{33j}, S_{33j}, R_{33j} *solving the following LMIs*

$$Q_{\mathscr{L}\pi} \prec 0, \quad Q_{\mathscr{L}\sigma} \prec 0, \tag{3.32a}$$

$$\begin{bmatrix} Q_{j\mathscr{L}} & M_{12} & M_{13} & M_{14} \\ * & M_{22} & B_j^T P_j E_{\pi j} & B_j^T P_j E_{\sigma j} \\ * & * & M_{33} & 0 \\ * & * & * & M_{44} \end{bmatrix} \prec 0, \tag{3.32b}$$

$$P_j \succ 0, \quad Q_{11j} \prec 0, \quad R_{11j} \succ 0, \tag{3.32c}$$

$$j = 1 \ldots h,$$

where

$$M_{12} = A_j^T P_j B_j - C_j^T S_{11j}, \quad M_{13} = A_j^T P_j E_{\pi j} - F_{\pi j}^T \Psi_{w_j}^T S_{22j} \Psi_{v_j},$$

$$M_{14} = A_j^T P_j E_{\sigma j} - F_{\sigma j}^T S_{33j}, \quad M_{22} = B_j^T P_j B_j - R_{11j},$$

$$M_{33} = E_{\pi j}^T P_j E_{\pi j} - \Psi_{v_j}^T R_{22j} \Psi_{v_j}, \quad M_{44} = E_{\sigma j}^T P_j E_{\sigma j} - R_{33j},$$

$$Q_{j\mathscr{L}} = A_j^T P_j A_j - \tau_j P_j - C_j^T Q_{11j} C_j - F_{\sigma j}^T Q_{33j} F_{\sigma j} - F_{\pi j}^T \Psi_{w_j}^T Q_{22j} \Psi_{w_j} F_{\pi j},$$

$$Q_{\mathscr{L}\pi} = \Psi_w^T Q_{22} \Psi_w + \Psi_w^T H_{\pi\pi}^T R_{22} H_{\pi\pi} \Psi_w, + \Psi_w^T S_{22} H_{\pi\pi} \Psi_w + \Psi_w^T H_{\pi\pi}^T S_{22}^T \Psi_w,$$

$$Q_{\mathscr{L}\sigma} = Q_{33} + S_{33} H_\sigma + H_\sigma^T S_{33}^T + H_\sigma^T R_{33} H_\sigma,$$

the decoupled control sequences $\{u_i \in \mathbb{R}^{m_i}\}$, $i = 1, 2, \ldots, h_{\Sigma_\Delta}$, *are QDC-attractable and bounded by the respective convex constraints* (3.31);
Then, the unconstrained controlled system Σ_Δ (3.2) *is asymptotically attractive.*

Proof (a) *Quadratical Dissipativity of Unit* $\mathscr{G}_j$: by substituting the model of $\mathscr{G}_j$ (3.24) and the parallelised connections of (3.25) into the dissipation inequality of (3.30), it follows that $\mathscr{G}_j$ is quadratically dissipative w.r.t the supply rate (3.29) (i.e. with respect to the output and input pair of

$$y_{jD} = (y_j^T, (\Psi_{w_j} w_{\pi j})^T, w_{\sigma j}^T)^T, u_{jD} = (u_j^T, (\Psi_{v_j} v_{\pi j})^T, v_{\sigma j}^T)^T$$

if LMI (3.32b) holds (see, e.g., [17]). Thus (3.30) is verified for all (y_{jD}, u_{jD}).

(*b*) *Asymptotic attractivity*: Define $\xi_c := u^T Q_c u + 2y^T S_c^T u + y^T R_c y$, the sum of all subsystem supply rates ξ_i, $i = 1, 2, \ldots, h_{\Sigma_\Delta}$, as in (3.31). Using (3.30) and the diagonality of $Q_{11}, Q_{22}, Q_{33}, S_{11}, S_{22}, S_{33}, R_{11}, R_{22}, R_{33}$, we obtain for

$$V\big(x(k)\big) := \sum_{j=1}^{h} V_j(x_j), \;\; V(x^+) - \tau\, V(x) \leqslant \xi\big((y, w_{\pi\pi}, w_\sigma), (u, v_{\pi\pi}, v_\sigma)\big),$$

where $\xi\big((y, w_{\pi\pi}, w_\sigma), (u, v_{\pi\pi}, v_\sigma)\big) = (y^T Q_{11} y + 2y^T S_{11} u + u^T R_{11} u)$
$$+ (w_{\pi\pi}^T Q_{22} w_{\pi\pi} + 2w_{\pi\pi}^T S_{22} v_{\pi\pi} + v_{\pi\pi}^T R_{22} v_{\pi\pi})$$
$$+ (w_\sigma^T Q_{33} w_\sigma + 2w_\sigma^T S_{33} v_\sigma + v_\sigma^T R_{33} v_\sigma).$$

Thus, $V(x^+) \leqslant \max_j(\tau_j)\, V(x) + \max_j(\gamma_j)\, |\xi_c^-|$. Similar to Theorem 2.1, $\|x\| \to 0$ as $k \to +\infty$. ∎

We have shown that the unknown parallel splitting ratios are successfully dealt with by employing the common input and output vectors $(v_{\pi\pi j}(k), w_{\pi\pi j}(k))$ representing the sums of parallelised vectors in the supply rate of a unit. We call this the parallelised masking technique to alleviate the conservativeness that may be caused by the constant splitting ratios of the material or energy flows of dynamically coupled subsystems. The decentralised MPC algorithm using the APRC-based attractivity constraint in this parallel masking method will be identical to those in Chap. 2 for each subsystem.

3.4   Numerical Examples

The numerical examples have been studied in simulation with MATLAB and Yalmip toolboxes in two cases studies for the same interconnection system with different scenarios to illustrate the effectiveness of the presented approaches to parallelised systems. Only the ASPRC with the matrix annihilation is studied here. The following subsystem state realisation matrices are deployed for this example:

$$\dot{x}(t) = \begin{bmatrix} A_1 & & \\ & A_2 & \\ & & A_3 \end{bmatrix} x(t) + \begin{bmatrix} B_1 & & \\ & B_2 & \\ & & B_3 \end{bmatrix} u(t) + \begin{bmatrix} E_1 & & \\ & E_2 & \\ & & E_3 \end{bmatrix} v(t), \quad (3.33)$$

$$w(t) = \begin{bmatrix} F_1 & & \\ & F_2 & \\ & & F_3 \end{bmatrix} x(t), \quad v(t) = Hw(t), \quad\quad\quad\quad (3.34)$$

$$y(t) = \begin{bmatrix} C_1 & & \\ & C_2 & \\ & & C_3 \end{bmatrix} u(t), \quad\quad\quad\quad\quad\quad\quad (3.35)$$

where

$$A_1 = \begin{bmatrix} -1.4 & 0.3 & 0 \\ 0 & -1.8 & 1.5 \\ 0.1 & -2.7 & 1.06 \end{bmatrix}, B_1 = \begin{bmatrix} 0 \\ 0 \\ 1 \end{bmatrix}, E_1 = \begin{bmatrix} 5.1 \\ 0 \\ 3.8 \end{bmatrix},$$

$$A_2 = \begin{bmatrix} -0.76 & 0 & 0.25 \\ .48 & -0.56 & 0 \\ 0 & 0.2 & -0.34 \end{bmatrix}, B_2 = -\begin{bmatrix} 1 & 0 & 0 \\ 0 & 1 & 0 \\ 0 & 0 & 1 \end{bmatrix}, E_2 = \begin{bmatrix} 4.2 \\ 0 \\ 0 \end{bmatrix},$$

$$A_3 = \begin{bmatrix} -4.3 & 5.9 \\ -1.8 & 2.7 \end{bmatrix}, B_3 = \begin{bmatrix} 1 & 0 \\ 0 & 1 \end{bmatrix}, E_3 = \begin{bmatrix} 6.95 \\ 0 \end{bmatrix},$$

$$C_1 \begin{bmatrix} 1 & 0 & 0 \\ 0 & 1 & 0 \\ 0 & 0 & 1 \end{bmatrix}, F_1 = \begin{bmatrix} 1 & 0 & 0 \end{bmatrix}, C_2 = \begin{bmatrix} 1 & 0 & 0 \\ 0 & 1 & 0 \\ 0 & 0 & 1 \end{bmatrix}, F_2 = \begin{bmatrix} 0 & 0 & 1 \end{bmatrix},$$

$$C_3 = \begin{bmatrix} 1 & 0 \\ 0 & 1 \end{bmatrix}, F_3 = \begin{bmatrix} 0 & 1 & 0 \end{bmatrix}, H = \begin{bmatrix} 0 & 0 & 1 \\ 1 & 0 & 0 \\ 0 & 0 & 0 \end{bmatrix}.$$

The predictive horizons $N_i \equiv 6$, $i = 1, 2, 3$, have been chosen for all subsystems. The MPC weighting matrices are as follows:

$$\mathscr{R}_1 = \text{diag}\{0.2, 0.2\}, \quad \mathscr{Q}_1 = \text{diag}\{1, 1, 2\}, \quad \mathscr{R}_2 = \text{diag}\{0.2, 0.2, 0.2\},$$
$$\mathscr{Q}_2 = \text{diag}\{1.5, 2.4, 1.6\}, \quad \mathscr{R}_3 = \text{diag}\{0.5, 0.5\}, \quad \mathscr{Q}_3 = \text{diag}\{1, 3\}.$$

The constraints on the control inputs are:

$$\|u_1\| \leqslant 2, \quad \|u_2\| \leqslant 2, \quad \|u_3\| \leqslant 2.$$

The initial states are as follows:

$$x_1(0) = \begin{bmatrix} 8 \\ -7 \\ -8 \end{bmatrix}, \quad x_2(0) = \begin{bmatrix} -6 \\ 8 \\ -1.5 \end{bmatrix}, \quad x_3(0) = \begin{bmatrix} .55 \\ -1.4 \end{bmatrix}.$$

The open-loop system is unstable with the eigenvalues of the global A below:

$$1.1634 + 0.2051i, \quad 1.1634 - 0.2051i, \quad 0.5411 + 0.1302i,$$
$$0.5411 - 0.1302i, \quad 0.8628 + 0.3398i, \quad 0.8628 - 0.3398i,$$
$$0.7623 + 0.3412i, \quad \text{and } 0.7623 - 0.3412i.$$

The decentralised MPC for the unconstrained system is not able to stabilise the interconnected system, as shown in Fig. 3.5. The simulation results with parallelised subsystems are provided here and compared to the results without parallel splitting connections. Consider three units of $\mathscr{G}_1$, $\mathscr{G}_2$ and $\mathscr{G}_2$ being interconnected as in Fig. 3.1. The unit state-space models are as follows:

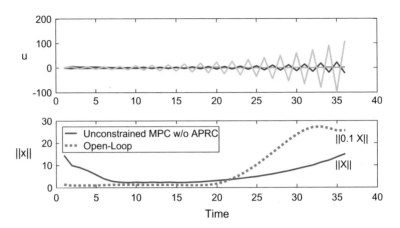

Fig. 3.5 Decentralised MPC without any constraints

$$\mathscr{G}_1 : A_1 = \text{diag}\left(\begin{bmatrix} -1.4 & .3 & 0 \\ 0 & -1.8 & 1.5 \\ .1 & -2.7 & 1.06 \end{bmatrix}, \begin{bmatrix} -1.6 & .2 & 0 \\ 0 & -2.1 & 1.7 \\ .3 & -1.8 & .9 \end{bmatrix}\right),$$

$$B_1 = \text{diag}[(0,0,1)^T, (0,0,1)^T], \ E_{\Delta 1} = \text{diag}[(2,0,2)^T, (2,0,2)^T], \ E_{\Theta 1} = 0,$$

$$C_1 = I_6, \ F_{\Delta 1} = \begin{bmatrix} 1 & 0 & 0 & 1 & 0 & 0 \end{bmatrix}, \ F_{\Theta 1} = \begin{bmatrix} 0 & 0 & 0 \end{bmatrix},$$

$$\mathscr{G}_2 : A_2 = \begin{bmatrix} -.76 & 0 & .25 \\ .48 & -.56 & 0 \\ 0 & .2 & -.34 \end{bmatrix}, \ B_2 = -\begin{bmatrix} 1 & 0 & 0 \\ 0 & 1 & 0 \\ 0 & 0 & 1 \end{bmatrix}, \ E_{\Delta 2} = 0, \ E_{\Theta 2} = \begin{bmatrix} 3.6 \\ 0 \\ 0 \end{bmatrix},$$

$$C_2 = I_3, \ F_{\Delta 2} = \begin{bmatrix} 0 & 0 & 0 \end{bmatrix}, \ F_{\Theta 2} = \begin{bmatrix} 0 & 0 & 1 \end{bmatrix},$$

$$\mathscr{G}_3 : A_3 = \begin{bmatrix} -4.3 & 5.9 \\ -1.8 & 2.7 \end{bmatrix}, \ B_3 = \begin{bmatrix} 1 & 0 \\ 0 & 1 \end{bmatrix}, \ E_{\Delta 3} = 0, \ E_{\Theta} = \begin{bmatrix} 6 \\ 0 \end{bmatrix},$$

$$C_3 = I_2, \ F_{\Delta 3} = \begin{bmatrix} 1 & 0 \end{bmatrix}, \ F_{\Theta 3} = \begin{bmatrix} 0 & 0 \end{bmatrix},$$

$$H_{\Delta} = \begin{bmatrix} 0 & 0 & 1 & 0 \\ 0 & 0 & 0 & 1 \\ 1 & 1 & 0 & 0 \end{bmatrix}, \quad H_{\Theta} = \begin{bmatrix} 0 & 0 & 0 & 0 \\ 0 & 0 & 0 & 0 \\ 0 & 0 & 0 & 0 \\ 0 & 0 & 1 & 0 \end{bmatrix}.$$

The predictive horizons are set with $N_1 = N_2 = N_3 = 6$. The weighting matrices of

$$\mathscr{Q}_1 = \text{diag}\{1, 1, 2; 1, 1, 1.2\}, \ \mathscr{R}_1 = \text{diag}\{2, 0.6\}, \ \mathscr{Q}_2 = \text{diag}\{1.5, 2.4, 1.6\},$$

$$\mathscr{R}_2 = \text{diag}\{0.5, 0.5, 0.5\}, \ \mathscr{Q}_3 = \text{diag}\{1, 3\}, \ \mathscr{R}_3 = \text{diag}\{0.5, 1\}, \text{ are chosen,}$$

and the initial state vectors of

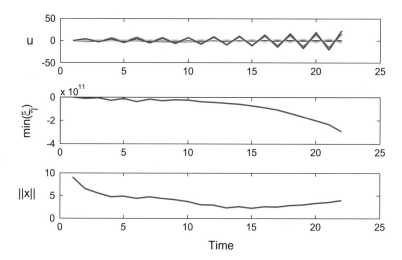

Fig. 3.6 Parallel splitting systems—decentralised MPC without ASPRCs and control constraints

$$x_1(0) = [2 \ -2 \ -3 \ -2 \ 1.2 \ -0.5]^T, \ x_2(0) = [-3 \ 2 \ -1]^T, \ x_3(0) = [1 \ -0.8]^T$$

are initialised. The following control constraints are considered in this simulation study: $\|u_{1_1}\| \leqslant 2, \ \|u_{1_2}\| \leqslant 2, \ \|u_2\| \leqslant 2, \ \|u_3\| \leqslant 2$.

3.4.1 Decentralised MPC Without Control Constraint

In this example, the decentralised MPC without ASPRCs is unable to stabilise the system when the control constraints are absent, as shown in Fig. 3.6.

3.4.2 Decentralised MPC with Control Constraint

The trajectories produced by the decentralised MPC with control constraints, but without the ASPRCs, are given in Fig. 3.7. The system is apparently not attractive. It also shows, in the same figure, that the global system is open-loop unstable via divergence trajectories. The value of $\|x\|$ is printed in this and the following figures, instead of $\|x\|^2$.

Now, a random variable is deployed to simulate the unknown splitting ratios, as follows:

$$E_{\pi 1} = \text{diag}\{\alpha(k)*2*(2, 0, 2)^T, \ (1-\alpha(k))*2*(2, 0, 2)^T\}, \ \alpha(k) \in (0, 1). \quad (3.36)$$

The trajectories for two different sets of $\alpha(k)$ are given in Figs. 3.8 and 3.9.

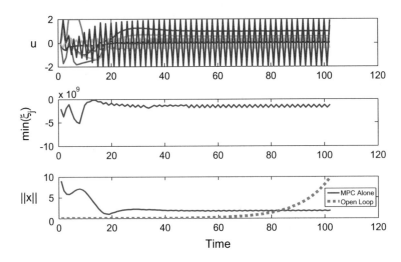

Fig. 3.7 Parallel splitting systems—decentralised MPC without ASPRCs

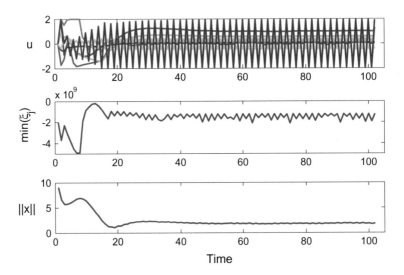

Fig. 3.8 Mixed connections—decentralised MPC without ASPRCs and unknown splitting ratios—simulation 1

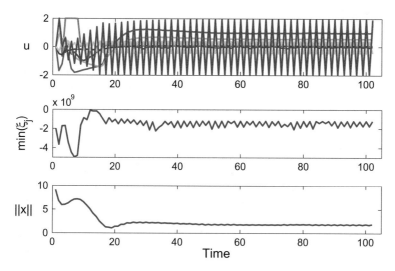

Fig. 3.9 Mixed connections—decentralised MPC without ASPRCs and unknown splitting ratios—simulation 2

3.4.3 Decentralised MPC with Control Constraint and ASPRC

When the decentralised MPC with ASPRCs is deployed, the trajectories of $u_j(k)$, $\|x(k)\|$ and $\min_{j=1,2,3} \xi_j$ are given in Fig. 3.10. The off-line computed coefficient matrices Q_{ji}, S_{ji}, R_{ji} are listed hereafter with the notation

$$R_j = \text{diag}[R_{ji}]_{i=1}^{g_j}, \ S_j = \text{diag}[S_{ji}]_{i=1}^{g_j}, \ Q_j = \text{diag}[Q_{ji}]_{i=1}^{g_j}$$

as per Theorem 3.1. The ASPRC coefficients $\gamma_1 = \gamma_2 = \gamma_3 = 0.99$ have been set in this numerical example (Fig. 3.11).

$$R_1 = -10^2 \begin{bmatrix} -1.2689 & -1.4172 & -0.0795 \\ -1.4172 & 9.7890 & -3.6063 \\ -0.0795 & -3.6063 & -3.8168 \end{bmatrix}, \ R_2 = -10^2 \begin{bmatrix} -2.2824 & -0.2336 & -0.0036 \\ -0.2336 & -1.7564 & -0.1088 \\ -0.0036 & -0.1088 & 9.2822 \end{bmatrix},$$

$$R_3 = -10^2 \begin{bmatrix} 6.8503 & 0.1993 \\ 0.1993 & -2.9051 \end{bmatrix},$$

$$S_1 = -10 \begin{bmatrix} 2.4649 \\ 0.7294 \\ -5.0120 \end{bmatrix}, \ S_2 = -10 \begin{bmatrix} 0.0012 & 0.1490 & 0.0146 \\ 0.1490 & 4.5902 & 0.2324 \\ 0.0146 & 0.2324 & 6.5412 \end{bmatrix},$$

$$S_3 = -10 \begin{bmatrix} 0.0018 & 0.1259 \\ 0.1259 & -1.5170 \end{bmatrix},$$

$$Q_1 = 3.0399 \, 10^2, \ Q_2 = -10^2 \begin{bmatrix} -2.2881 & 0.0003 & -0.0006 \\ 0.0003 & -2.4092 & -0.0030 \\ -0.0006 & -0.0030 & -2.4433 \end{bmatrix},$$

$$Q_3 = -10^2 \begin{bmatrix} -2.2744 & 0.0043 \\ 0.0043 & -2.3146 \end{bmatrix}.$$

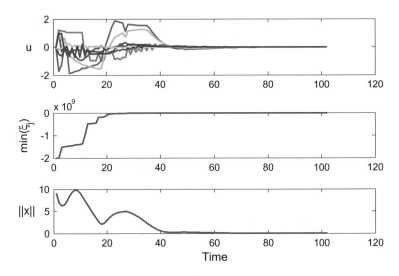

Fig. 3.10 Mixed connections—decentralised MPC with ASPRCs—equal splitting ratio

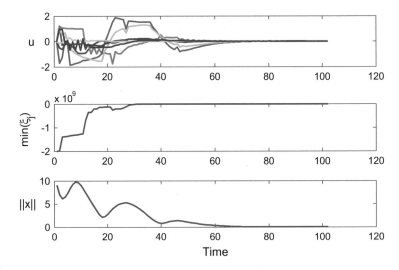

Fig. 3.11 Mixed connections—decentralised MPC with ASPRCs and unknown parallel splitting ratios

3.5   Concluding Remarks

The asymptotic attractivity conditions for interconnected systems with serial and parallelised connections have been derived in this chapter using a matrix annihilation from the null base of a block matrix. The parallelised connection structure is cancelled out, as a result of the matrix annihilation. Also, the asymptotic surely positive realness constraint (ASPRC), an enhanced version of the APRC, has been applied to the decentralised MPC scheme as an attractivity constraint for each unit.

Then, the parallelised connections with unknown parallel splitting ratios have been explicitly modelled and masked out, as an alternative, later in this chapter.

Numerical examples are simulated with MATLAB and Yalmip toolboxes for an interconnected system in two case studies to demonstrated the effectiveness of the presented decentralised MPC scheme with quadratic constraints for interconnected systems having mixed connection structures of serial and parallel splitting types.

The parallelised subsystems represent the parallel-redundant process streams of a unit operation in a large processing plant, or the control areas of a power system connected in parallel, and other systems.

Chapter 4
Quadratic Constraint for Semi-automatic Control

The quadratic constraint, APRC or QDC, is implemented separately from the control algorithm in this chapter instead of integrating with the MPC optimisation as in previous chapters. The real-time value of the supply rate is monitored by an independent entity, called *stabilising agent*, while the quadratic constraint is continuously checked online by the associated semi-automatic control system. The stabilising agent will override the control actions of the controller, or of the human operator, if the quadratic constraint is violated, thus confine the closed-loop system within the QDC ellipsoidal boundary − a stabilising region. The closed-loop system will be asymptotically attractive, or stabilisable in the GDC sense, as a result of these overriding actions online.

4.1 Semi-automatic Control

In the processing industries, a complex plant can be viewed as a larger scale global system formed by numerous subsystems interconnected to each other. By virtue of the intricate interactions between state variables within a subsystem, the computerised-control system is normally designed such that a dedicated multi-variable controller is installed for each of its single subsystem.

These multiple installations create new requirements on the orchestration of multi-variable controllers to avoid a possible destabilisation owing to the coupling effects between subsystems. To assure the continuous operability and maintainability of a complex plant, these multi-variable controllers usually operate in two modes, automatic and manual. The manual mode allows for human–machine interfacing via the human–machine interface (HMI) functions and switching between the remote control room operator and the automatic control algorithm, as illustrated in Fig. 4.1. The global system thus operates semi-automatically with the two control modes, feed-

© Springer Nature Singapore Pte Ltd. 2018
A. Tri Tran C. and Q. Ha, *A Quadratic Constraint Approach to Model Predictive Control of Interconnected Systems*, Studies in Systems, Decision and Control 148, https://doi.org/10.1007/978-981-10-8409-6_4

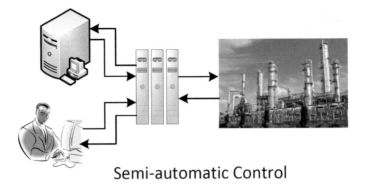

Fig. 4.1 Semi-automatic control with human operator

Fig. 4.2 Semi-automatic
control with stabilising
agent. Among the control
vectors in this figure, $\hat{u}$ is the
planned control vector while
$\tilde{u}$ is the stabilisable control
vector

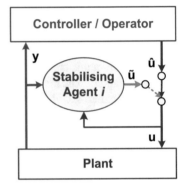

back control and manual control. The semi-automatic control operations may consist
of subsystem start-up, warm restart after a shutdown, subsystem operational mode
changes, or simply control room operator actions. Confining the interconnected system within the stabilising boundary in different modes will thus be essential for the
safe and reliable operations.

A stabilising bound for the semi-automatic control of interconnected systems in a
decentralised architecture is developed in this chapter. Figure 4.2 shows a representation for the interactions between a semi-automatic control system and its associated
stabilising agent. The notion of 'stabilising agent' has been geminated for designing the stabilising mechanism in a semi-automatic control system. The quadratic
constraint, APRC or QDC, will be used by the stabilising agent to derive the stabilising bound for the semi-automatic control system independently to the automatic
controller. The method of stabilising an interconnected system with the stabilising
agents stems from the idea of splitting the stabilising mechanism with the control
algorithm, inspired by the segregation principle in the reliability designs; see, e.g.
[136, 163].

In brief, the stabilising agent analyses the information obtained from the system dynamics and then computes a stabilising bound that can prevent a possible destabilisation. The stabilising agent will initiate an overriding action to steer the system into the APRC-based (or QDC-based) stablisable region upon any deviations. In other words, the stabilising agent continuously monitors the system motion and limits it within the APRC-based (or QDC-based) stabilisable boundary. The segregation of a stabilising agent from its associated control system will ensure that the closed-loop and human-in-the-loop systems are stabilisable semi-automatically.

4.2 Stabilising Agent Operation

Consider the state-space model of an interconnected system having decoupled control constraints in Chap. 2. In this section, the MPC optimisation without the quadratic constraint is updated at every time step to control the corresponding subsystem. As a solution for the attractivity assurance, the control inputs will be event-based corrected in a segregated manner by the stabilising agent such that the quadratic constraint is always satisfied. In other words, we are concerned with the design of h disparate stabilising agents for h associated subsystems $\mathscr{S}_i, i = 1, 2, \ldots, h$, in a decentralised control architecture. A stabilising agent will read the input and output vectors of $\mathscr{S}_i$, which can be in the automatic control mode with MPC or in the manual control mode with the operator. Their current-time and historical data will then be used to deduce the APRC-based (or QDC-based) stabilising bound for the inputs of $\mathscr{S}_i$. The decentralised stabilising agents for an interconnected system are shown in Fig. 4.3. The evolution of the stabilising bound in two consecutive time steps is represented by Fig. 4.4.

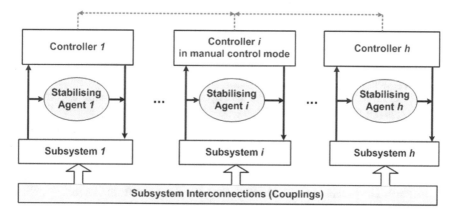

Fig. 4.3 Decentralised stabilising agents for an interconnected system

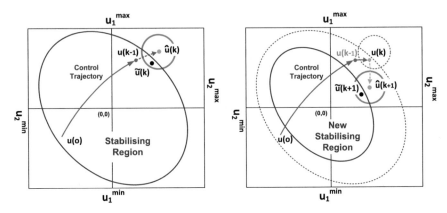

Fig. 4.4 Evolution of stabilising bound and adjustment of a stabilising agent in two consecutive time steps

The activities of a stabilising agent shown in Fig. 4.2 comprise the following:

1. Assuming the control vector generated from the MPC or from the human operator is $\hat{u}$, a stabilising agent will:—

2. Monitor the values of the supply rate $\xi_i(u_i(k), y_i(k))$ and check if it is negative. The controller runs its own control algorithm and manipulates $\mathscr{S}_i$ until the first initial time $k_0(i)$, at which $\xi_i(u_i(k_0), y_i(k_0)) < 0$ is detected. If a tuning scalar ρ_i is used to prevent the conservative actions, the automatic controller will manipulate $\mathscr{S}_i$ until the first initial time $k_0(i)$, at which $\xi_i(u_i(k_0), y_i(k_0)) < -\rho_i^2$ is detected, instead of $\xi_i(u_i(k_0), y_i(k_0)) < 0$.

3. Compute the stabilising bound $\tilde{u}_i(k)$ whenever $\xi_i(u_i(k_0), y_i(k_0)) < 0$ or $\xi_i(u_i(k), y_i(k)) < -\rho_i^2$ from the quadratic constraint (2.27) and the control constraint (2.2).

4. If the ellipsoidal region governed by the quadratic constraint (2.27) does not intersect the control constraint (2.2), update the supply function coefficient matrices Q_i, S_i, R_i of the quadratic constraint (2.27) from the calculated $\xi_i(u_i(k-1),$ $y_i(k-1))$ and the known value of $y(k)$ (refer to the stabilisability conditions in Chap. 2 and other chapters).

5. Override the control $\hat{u}_i$ from MPC with $\tilde{u}_i(k)$ if the updated quadratic constraint (2.27) is not fulfilled by $\hat{u}_i$.

6. The stabilising agent will be active until the time instant $\bar{k} > k_0$, at which $\xi_i(u_i(\bar{k}), y_i(\bar{k})) \geqslant 0$ while $\xi_i(u_i(\bar{k}-1), y_i(\bar{k}-1)) < 0$. The stabilising agent will then be inactive if the convex quadratic constraint (2.27) becomes concave in the case of using the APRC (this time instant $\bar{k}$ does not exist when the QDC is used). The process is rolled back to Step 1 and repeats from thereon.

In short, the stabilising agent initially remains inactive. It only monitors $\xi_i(k)$ until the first time instant $k_0(i)$, at which $\xi_i < -\rho_i^2$. The time step $k_0(i)$ then becomes the initial time for activating the procedure that computes the stabilising bound, updates the coefficient matrices of the quadratic constraint if necessary and overrides the control actions if the quadratic constraint is not fulfilled. It is noting, however, that the initial time k_0 can be chosen a priori as 0 in many applications.

The QDC-based stabilising agent activities in two consecutive time steps in Fig. 4.4 are interpreted as follows:

- The control box constraint is represented by the black rectangular with the maximum and minimum values u_1^{max}, u_1^{min}, u_2^{max} and u_2^{min}.
- The QDC ellipsoids are represented by the blue ellipses, at the time steps $k - 1$ on the left and k on the right-hand side.
- We assume that the executed control $u(k - 1)$ is inside the QDC ellipse at the time step k, as drawn in the left figure. The MPC provides the control move $\hat{u}(k)$ at the time step k, which lies outside the QDC ellipse, i.e. violates the QDC-based stabilising boundary. The desirable control move should be $\tilde{u}(k)$ inside the QDC ellipse instead. The stabilising agent thus overrides the MPC control move $\hat{u}(k)$ with the stabilising control move $\tilde{u}(k)$ to manipulate the plant.
- At the next time step $k + 1$, the already executed control $u(k)$ is outside the QDC ellipse at the time step $k + 1$, as drawn in the right figure. The MPC then provides a planned control move $\hat{u}(k + 1)$ at the time step $k + 1$, which also lies outside the QDC ellipse, i.e. violates the QDC-based stabilising boundary again. Since the desirable control move should be $\tilde{u}(k + 1)$ inside the QDC ellipse, the stabilising agent overrides the MPC control move $\hat{u}(k + 1)$ with the stabilising control move $\tilde{u}(k + 1)$ to drive the plant.

The detailed algorithm of a stabilising agent is outlined in the next subsection.

4.3 Constructive Procedure for Stabilising Agents

Assuming that the sampling and updating time instants of all subsystems and their controllers are synchronised, the detailed algorithm of a stabilising agent is given in Procedure 4.1 below.

4.3.1 Stabilising Agent Procedure

Procedure 4.1 Stabilising agent for a subsystem $\mathscr{S}_i$.

1. **Off-line**:

    - The dissipativity matrices X_i, Y_i, Z_i, P_i and coefficient matrices Q_i, S_i and R_i of the supply rate are predetermined from the LMI optimisation.
    - Select the tuning parameter ρ_i.
    - Select γ_i as a tuning parameter for the quadratic constraint.

2. **Online**: At every time step $k > 0$, the control $\hat{u}_i(k)$ is computed by the local MPC or set by the operator, and then the quadratic constraint of $\mathscr{S}_i$ is:

- *Step* 1: Inactive until $k_0(i)$.
  a. Calculate $\xi_i(u_i(k), y_i(k))$ from $\hat{u}_i(k)$ and $y_i(k)$ using the multiplier matrices Q_i, S_i and R_i;
  b. Verify $\xi_i(\hat{u}_i(k), y_i(k)) < -\rho_i^2$. If true, assign k to $k_0(i)$, and go to Step 2, otherwise go back to 1(a);
- *Step* 2: Active with the stabilising bound computation.
  a. Whenever $\xi_{ic}(\hat{u}_i(k), y_i(k)) < \gamma_i \times \xi_{ic}(u_i(k-1), y_i(k-1))$,
     i. Update the multiplier matrices Q_i, S_i, R_i, if necessary, such that one of the control qualification conditions in Sect. 2.2.2 in Chap. 2 is fulfilled;
     ii. Compute the stabilising bound vector $\tilde{u}_i(k)$ by solving the optimisation problem of the following:

$$\min_{\tilde{u}_i(k)} \|\tilde{u}_i(k) - \hat{u}_i(k)\|^2,$$

s.t. $\xi_{ic}(\tilde{u}_i(k), y_i(k)) \geqslant \gamma_i \times \xi_{ic}(\hat{u}_i(k), y_i(k))$ and $\| \tilde{u}_i(k) \|^2 \leqslant \eta_i$; (4.1)

   iii. Override $\hat{u}_i(k)$ with $\tilde{u}_i(k)$ and output as $u_i(k)$ to control $\mathscr{S}_i$;

- Verify $\xi_i(\hat{u}_i(k), y_i(k)) < -\rho_i^2$, if true go back to 2(a), otherwise assign k to $\bar{k}(i)$, deactivate the stabilising agent and go back to Step 1 above*.

Remark 4.1 The stabilising agent algorithm in Procedure 4.1 above is for the quadratic constraint that can become negative at several repeated time instants $k_0(k)$. As a result, this algorithm is applicable to the REDC-type quadratic constraint defined in Chap. 2.

4.3.2 Graphical Presentation

Consider a subsystem having two control inputs, i.e. u_i has two elements, $u_i^{(1)}$ and $u_i^{(2)}$. The control plane plot with two axes of $u_i^{(1)}$ and $u_i^{(2)}$ is shown in Fig. 4.5. The plot in this figure consists of the control trajectory, in red, of $u_i(j)$, the quadratic constraint ellipsoidal (stabilisable) regions, the planned control $\hat{u}_i$, the balls of $\|\tilde{u}_i(k) - \hat{u}_i(k)\|^2 \leqslant \varepsilon$, in a green circle, and the stabilising bound $\tilde{u}_i$ from the solution to the optimisation problem (4.1) in Procedure 4.1.

The two cases of incremental changes of the supply rate ξ_i are included in this figure. Case (a) shows the convex region of $0 \geqslant \xi_i \geqslant \gamma_i \xi_i(k-1)$ which is the main focus of the stabilising agent in this chapter. Case (b) shows the region of $0 < \xi_i \leqslant \xi_i(k-1)$ with the APRC (when the QDC is used, Case (b) does not exists).

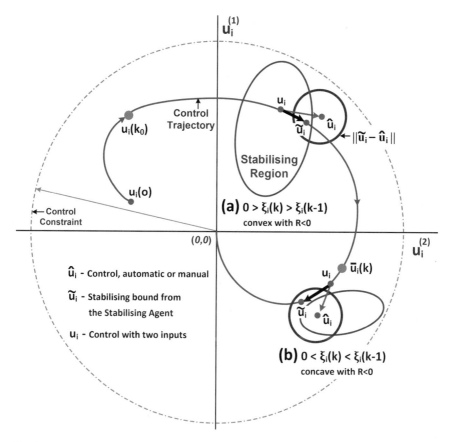

Fig. 4.5 Control plane plot with ellipsoids of quadratic constraints and the stabilising bound calculations

After the point $u_i(k_0)$ (and before the point $u_i(\bar{k})$) along the control trajectory, $\xi_i < 0$. The control u_i represents $u_i(k-1)$. Assume that the control algorithm, or the operator, generates the planned $\hat{u}_i(k)$ outside the ellipsoidal region in Case (a). The stabilising agent then solves the problem (4.1) for $\tilde{u}_i(k)$ which is represented by a point inside a ball taking $\hat{u}_i(k)$ as the centre. The stabilising bound $\tilde{u}_i(k)$ is inside the ellipsoidal region for the convex Case (a), and reversely for Case (b) if the APRC is used, and also inside the ball $\|\tilde{u}_i(k) - \hat{u}_i(k)\|^2 \leqslant \varepsilon$ in both cases.

The two steps of Procedure 4.1 correspond to the first two sections of the control trajectory. The initial Step 1 takes place between $u_i(0)$ and $u_i(k_0)$. The main activity of a stabilising agent is in Step 2, between the two points $u_i(k_0)$ and $u_i(\bar{k})$ for the case of APRC (when the QDC is used, the point $u_i(\bar{k})$ does not exists). Beyond the point $u_i(\bar{k})$, the APRC is deactivated and repeats the polling for $\xi_i < -\rho_i^2$. The constraint of the form $\xi_i \leqslant \xi_i(k-1)$ is concave, with $R \prec 0$, as shown in Fig. 4.5. The controller should stabilise the system beyond $\bar{k}$, or the process will restart from a new k_0 upon the recurrence of $\xi_i < -\rho_i^2$. If this condition re-emerges, the constraint of the REDC type defined in Chap. 2 becomes applicable, but not the APRC, accordingly.

The development for nonzero set points and systems with disturbances is given in the next section.

4.4 Stabilising Agent with Output Tracking

This section is concerned with the problem of output tracking for interconnected systems with input disturbances, as shown in Fig. 4.6, by using stabilising agents.

Consider Σ (2.1) consisting of h subsystems with input disturbances as follows:

$$
\mathscr{S}_i : \begin{cases}
x_i(k+1) = A_i x_i(k) + B_i u_i(k) + E_i v_i(k) + J_i d_i(k), \\
y_i(k) = C_i x_i(k), \\
w_i(k) = F_i x_i(k),
\end{cases}
\tag{4.2}
$$

where $u_i \in \mathbb{U}_i \subset \mathbb{R}^{m_i}$, $y_i \in \mathbb{R}^{q_i}$, $x_i \in \mathbb{R}^{n_i}$, $v_i \in \mathbb{R}^{m_{v_i}}$, $w_i \in \mathbb{R}^{q_{w_i}}$; x_i, u_i, y_i, v_i, and w_i are, respectively, state, control, measurement output, interactive input and interactive output vectors; d_i represents the unknown but bounded state/input disturbance, $d_i \in \mathbb{R}^{m_i}$. In this section, the following control constraint is considered:

$$
\mathbb{U}_i = \{u_i : \|u_i - \bar{u}_i\|^2 \leqslant \eta_i\},
\tag{4.3}
$$

where $\bar{u}_i$ is the nonzero steady state of $u_i(k)$.

Similarly to models in Chap. 2, the larger scale global system Σ is represented by the state-space model with a block-diagonal system $\mathscr{S}$ formed by h subsystems $\mathscr{S}_i$ and the global interconnection process H, as follows:

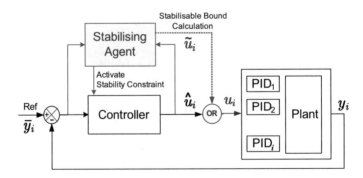

Fig. 4.6 Stabilising agent with output tracking

$$\Sigma : \begin{cases} x(k+1) = A\,x(k) + B\,u(k) + E\,v(k) + J\,d(k), \\ \quad y(k) = C\,x(k), \\ \quad w(k) = F\,x(k), \\ \quad v(k) = H\,w(k). \end{cases} \tag{4.4}$$

From a centralised point of view, Σ is expressed as

$$\Sigma : \begin{cases} x(k+1) = A_\Sigma\,x(k) + B\,u(k) + J\,d(k), \\ \quad y(k) = C\,x(k), \end{cases} \tag{4.5}$$

where $A_\Sigma = A + EHF$, with nonsingular $I - A_\Sigma$ for the output-tracking control problem.

When all the interactive inputs and outputs vanish (i.e. $v_i = 0$ and $w_i = 0$), a stand-alone subsystem $\mathscr{S}_i$, denoted as $\mathscr{S}_i^s$, has the state-space model of the form:

$$\mathscr{S}_i^s : \begin{cases} x_i(k+1) = A_i x_i(k) + B_i u_i(k), \\ \quad y_i(k) = C_i x_i(k). \end{cases} \tag{4.6}$$

Assumption: (1) The state-space realisations (A_i, B_i) in (4.6) and (A_Σ, B) in (4.5) are controllable, and also, (A_i, C_i) and (A, C) are observable. (2). The matrix $[(I - A_\Sigma)^T \ C^T]^T$ is full-rank.

The stability condition for the output-tracking problem is based on a combination of the steady-state-independent quadratic constraint and the dissipation inequality around the operating point. Denote the output-tracking input and output vectors as follows:

$$u_i^{ot}(k) = u_i(k) - \bar{u}_i, \quad y_i^{ot}(k) = y_i(k) - \bar{y}_i,$$

where $(\bar{u}_i, \bar{y}_i)$ is the steady-state pair. The following norm-bounded gain for the disturbance d_i is considered for the controller design:

$$\|d_i(k)\| \leqslant \gamma_i \, \|y_i^{ot}(k)\|, \tag{4.7}$$

in which the gain γ_i is constant and given.

In this chapter, the quadratic constraint coefficient matrices Q, S, R are determined online in conjunction with LMIs in the asymptotic attractivity condition stated in Theorem 4.1 below. The steady-state-independent quadratic constraint is first introduced below.

4.4.1 Steady-State-Independent Quadratic Constraint

Define the following steady-state-independent quadratic supply rate function for $\mathscr{S}_i$:

$$\xi_{ic}\big(u_i^{ot}(k),\, y_i^{ot}(k)\big) := \begin{bmatrix} u_i^{ot}(k) \\ y_i^{ot}(k) \end{bmatrix}^T \begin{bmatrix} Q_{ic} & S_{ic} \\ S_{ic}^T & R_{ic} \end{bmatrix} \begin{bmatrix} u_i^{ot}(k) \\ y_i^{ot}(k) \end{bmatrix}, \tag{4.8}$$

where Q_{ic}, S_{ic}, R_{ic} are the coefficient matrices with symmetric Q_{ic} and R_{ic}.

For conciseness, $\xi_{ic}\big(u_i^{ot}(k),\, y_i^{ot}(k)\big)$ is denoted by ξ_{ic}^k, and the time index k is represented by a superscript typeface k without brackets, or omitted, where appropriate. It is assumed in this chapter that there exists an initial time $k_0 \leqslant 0$ such that $\xi_{ic}^{k_0} < 0$.

Definition 4.1 The controlled motion $\big(u_i^{ot}(k),\, y_i^{ot}(k)\big)$ is said to satisfy the asymptotically positive realness constraint with respect to ξ_{ic}^k (4.8) for the output-tracking problem, or simply APRCO, if there are $k_0 \in \mathbb{Z}^+$, $0 \leqslant \zeta_i < 1$, $\varepsilon = 1 - \zeta_i$ and $|\beta_i^k| \geqslant \sqrt{\varepsilon \times |\xi_{ic}^{k-1}|}$, such that

$$\xi_{ic}^k \geqslant \frac{1}{2}\, \sigma_i^k \times \big(\xi_{ic}^{k-1} + (\sigma_i^k - 1) \times (\beta_i^k)^2\big) \quad \forall k > k_0, \tag{4.9}$$

where $\sigma_i^k = 1 - \mathsf{sgn}(\xi_{ic}^{k-1})$.

In the above definition, the bias number β_i^k should be sufficiently small to ensure that a smooth control trajectory is generated by the stabilising agent.

If $\xi_{ic}^{k-1} < 0$ and $\xi_{ic}^k < 0$, the above APRCO can be rewritten as

$$\xi_{ic}^k \geqslant \xi_{ic}^{k-1} + (\beta_i^k)^2 \quad \forall k > k_0. \tag{4.10}$$

Alternatively, the following practical APRCO can also be considered:

$$\xi_{ic}^k \geqslant \gamma_i \times \xi_{ic}^{k-1} - \lambda_i \quad \forall k > k_0, \tag{4.11}$$

with $\lambda_i > 0$ and $0 < \gamma_i < 1$.

It is straightforward to see that if ξ_{ic}^k satisfies the APRCO (4.9) whenever $k > k_0$, then for every real number $\delta_i \geqslant 0$, there exists $\bar{k}_i \in \mathbb{Z}^+$ such that

$$\xi_{ic}^k \geqslant -\delta_i \quad \forall k \geqslant \bar{k}_i. \tag{4.12}$$

Now, by defining

$$Q_c := \mathsf{diag}[\,Q_{ic}\,]_1^h,\quad S_c := \mathsf{diag}[\,S_{ic}\,]_1^h,\quad R_c := \mathsf{diag}[\,R_{ic}\,]_1^h,\quad W_c := \mathsf{diag}[\,W_{ic}\,]_1^h,$$

and

$$\xi_{dc}^k := u_{ot}^{kT} Q_c u_{ot}^k + 2 y_{ot}^{kT} S_c u_{ot}^k + y_{ot}^{kT} R_c y_{ot}^k + d^{kT} W_c d^k,$$

as the supply rate of the larger scale global system Σ ($u \in \mathbb{R}^m$), where

$$u_{ot}^k := u^k - \bar{u},\quad y_{ot}^k := y^k - \bar{y},\quad \bar{u}^T := [\bar{u}_i^T \ \dots \ \bar{u}_i^T],\quad \bar{y}^T := [\bar{y}_i^T \ \dots \ \bar{y}_i^T],$$

the convergence condition is stated in the next theorem.

4.4.2 Convergence Condition with Output Tracking

The system motions are required to be bounded by the APRC as defined in (4.9), in conjunction with the quadratical dissipativity of Σ with respect to the supply rate ξ^k of the form

$$\xi^k := y_{ot}^{kT} Q y_{ot}^k + 2 y_{ot}^{kT} S u_{ot}^{kT} + u_{ot}^{kT} R u_{ot}^k + d_k^T W d_k, \qquad (4.13)$$

where Q, R, W are symmetric matrices, and

$$Q := \mathrm{diag}[Q_i]_1^h, \quad S := \mathrm{diag}[S_i]_1^h, \quad R := \mathrm{diag}[R_i]_1^h, \quad W := \mathrm{diag}[W_i]_1^h.$$

Denote $x_{ot}(k) := x(k) - \bar{x}$, where $\bar{x}$ is the steady state of $x(k)$. The storage function of the form

$$V(x_{ot}(k)) = x_{ot}^T(k) P x_{ot}(k), \quad P = P^T \succ 0,$$

is considered herein.

Theorem 4.1 Let $0 \leqslant \lambda_i < 1, 0 < \tau < 1$ and $k_0 \geqslant 0$. Suppose that
  (i) The following LMIs be feasible in P, Q_i, S_i, R_i, W_i, Q_{ic}, S_{ic}, R_{ic}, W_{ic}:

$$\begin{bmatrix} P & A_\Sigma^T P & B^T P & J^T P & 0 \\ * & \tau P + C^T Q C & C^T S & 0 & M_{25} \\ * & * & R & 0 & M_{35} \\ * & * & * & W & M_{45} \\ * & * & * & * & M_{55} \end{bmatrix} \succcurlyeq 0, \ P \succ 0, \ R \succ 0, \ W \succ 0, \ (4.14)$$

$$\begin{bmatrix} Q_i & S_i + S_{ic} \\ S_i^T + S_{ic}^T & R_i + Q_{ic} \end{bmatrix} \prec 0, \text{ and } W_{ic} + W_i \prec 0, \qquad (4.15)$$

$$i = 1 \ldots h,$$

where

$$M_{25} := (I - A_\Sigma)^T P \Pi - C^T Q C \Pi - C^T S, \quad \Pi := (I - A_\Sigma)^{-1} B,$$
$$M_{35} := -B^T P \Pi - S^T C \Pi - R, \quad M_{45} := -J^T P \Pi, \quad M_{55} := R + \Pi^T C^T Q C \Pi;$$

  (ii) With feasible matrices Q_{ic}, W_{ic} and S_{ic} from the LMIs in (4.14) and (4.15), the following LMIs be feasible in R_{ic} for $i = 1, 2, \ldots, h$:

$$R_{ic} + \gamma_i^2 \|\hat{W}_{ic}\| I_{q_i \times h} \prec 0, \quad \hat{W}_{ic} = -W_{ic} \succ 0, \quad R_{ic} \prec 0, \qquad (4.16a)$$

$$y_i^{ot\,T} \left(R_{ic} - S_{ic}^T (Q_{ic})^{-1} S_{ic} \right) y_i^{ot} - \gamma_i \times \xi_{ic}^{k-1} > 0, \qquad (4.16b)$$

$$\begin{bmatrix} -\rho_i \times \eta_i - \gamma_i \times \xi_{ic}^{k-1} + y_i^{ot\,T} R_{ic}\, y_i^{ot} & y_i^{ot\,T} S_{ic} \\ S_{ic}^T y_i^{ot} & Q_{ic} + \rho_i I_{m_i} \end{bmatrix} \preccurlyeq 0; \qquad (4.16c)$$

(iii) There exist a compact set $\mathbb{X} \subset \mathbb{R}^n$, $\bar{x} \in \mathbb{X}$, *such that the system* Σ *(4.5) is APRCO attractability with some h control sequences* $\{u_i(k) \in \mathbb{U}_i\}$, $i = 1, 2, \ldots, h$ *whose each* $u_i(k)$ *satisfies the corresponding APRCO (4.9).*

Then, $x(k)$ *remains bounded and* $x(k) \to 0$ *as* $k \to \infty$ *with some of the APRCO-attractable control sequences* $\{u_i(k) \in \mathbb{U}_i\}$, $i = 1, 2, \ldots, h$.

Proof (1) *Output-tracking dissipation*: From (4.14), it follows that the global system Σ is dissipative w.r.t. the supply rate (4.13) around $\bar{y}$ and $\bar{u}$ (the dissipation inequality holds true for all $x, u, d, \bar{u}$). (2) *Convergence*: Under the condition (*ii*), there is unconstrained $u_i(k)$ feasible to (4.9). Thus, for each $\varepsilon_i > 0$, there always exists $\bar{k}_i$ such that $0 > \xi_{ic}^{\bar{k}_i - 1} \geqslant -\varepsilon_i$. Thus, the constraint $0 > \xi_c(k) \geqslant -\varepsilon \, \forall k > \bar{k} > k_0$, where $\bar{k} := \sup_{i=1 \ldots h} \bar{k}_i$ and $\varepsilon := \max_i \varepsilon_i$, is realisable with a finite $u_i(k)$, $i = 1, 2 \ldots, h$. By the dissipativity of Σ which is verified by (4.14) and the satisfaction of APRCO of each subsystem, it follows that

$$V(x_{ot}^{k+1}) - \tau \times V(x_{ot}^k) + \xi_{dc}^{k-1} \leqslant \xi^k + \xi_{dc}^k - d^{kT} W_c d^k \quad \forall k > \bar{k},$$

$$\xi^k + \xi_{dc}^k = (*) \begin{bmatrix} Q + R_c^k & S + S_c \\ * & R + Q_c \end{bmatrix} \begin{bmatrix} y_{ot}^k \\ u_{ot}^k \end{bmatrix} + (*)(W + W_c)d^k.$$

Now, (4.7), (4.15) and (4.16a) result in $y_{ot}^{kT} R_{ic}^k y_{ot}^T - d_i^{kT} W_{ic} d^k \leqslant 0$ and $W_{ic} \prec 0$. Thus

$$V(x_{ot}^{k+1}) - \tau \times V(x_{ot}^k) + \gamma_i \xi_{dc}^{k-1} \leqslant -\tau_1^2(\|y_{ot}^k\| + \|u_{ot}^k\|) - \tau_2^2\|d^k\|,$$

for $\tau_1 > 0$ and $\tau_2 > 0$. From the observability of Σ and the assumption (*ii*) on the full-rank of $[(I - A_\varepsilon)^T \ C^T]^T$, we obtain for a small $\eta > 0$,

$$V(x_{ot}^{k+1}) - \tau \times V(x_{ot}^k) \leqslant -x_{ot}^{kT}(\eta I)x_{ot}^k \ \forall k \geqslant \bar{k}. \qquad (4.17)$$

Due to $P \succ 0$, there are $0 < \nu_1 < \nu_2$ such that $\nu_1 x_{ot}^T x_{ot} \leqslant x_{ot}^T P x_{ot} \leqslant \nu_2 x_{ot}^T x_{ot}$. Therefore,

$$x_{ot}^T(k+1)Px_{ot}(k+1) \leqslant \tau \times (1 - \frac{\eta}{\nu_2})x_{ot}^T(k)Px_{ot}(k) \ \forall k \geqslant \bar{k}.$$

$$\Rightarrow x_{ot}^T(k)x_{ot}(k) \leqslant \tau \times \nu_2 \times \nu_1^{-1} \times \nu_3^{2(k-\bar{k})} x_{ot}^T(\bar{k})x_{ot}(\bar{k}), \text{ where } \nu_3^2 = \frac{\nu_2 - \eta}{\nu_2}, 0 < \nu_3 < 1$$

by choosing $1 > \nu_2 > \eta$. Hence, $\|x_{ot}(k)\| \leqslant \gamma \nu_3^{(k-\bar{k})}\|x_{ot}(\bar{k})\|, \gamma = \sqrt{\tau \times \nu_2 \times \nu_1^{-1}}$.

From the above inequality, for a given $\varepsilon > 0$, let $\delta := \gamma^{-1}\varepsilon$, and since $0 < \nu_3 < 1$, $\|x_{ot}(\bar{k})\| < \delta \implies \|x_{ot}(k)\| < \varepsilon \; \forall k \geqslant \bar{k}$.

(3) *Control qualification condition for APRCO*: By rewriting $\xi_{ic}(u_i^{ot}, y_i^{ot})$ as

$$\xi_{ic}(u_i^{ot}, y_i^{ot}) = -(u_i^{ot} - (Q_{ic})^{-1} S_{ic} y_i^{ot})^T (-Q_{ic})(u_i^{ot} - (Q_{ic})^{-1} S_{ic} y_i^{ot})$$
$$+ y_i^{ot^T} R_{ic}^k y_i^{ot} - y_i^{ot^T} S_{ic}^T (Q_{ic})^{-1} S_{ic} y_i^{ot},$$

and since $Q_{ic} \prec 0$ (the second LMI in (4.15)), the feasible region is an ellipsoid w.r.t u_i^{ot} which is not shrinking to a point by virtue of (4.16b) and $|\beta_i^k| \geqslant \sqrt{\varepsilon_i \times |\xi_{ic}^{k-1}|}$ (4.9), and a subset of $\mathbb{U}_i$ (4.3) if for $\rho_i > 0$, the following inequality is fulfilled for every $u_i^{ot}(k)$:

$$\rho_i \times \left[u_i^{ot^T} I u_i^{ot} - \eta_i \right] - \left[u_i^{ot^T}(-Q_{ic})u_i^{ot} + 2y_i^{ot^T}(-S_{ic})u_i^{ot} \right.$$
$$\left. + y_i^{ot^T}(-R_{ic}^k)y_i^{ot} + \gamma_i \times \xi_{ic}^{k-1} \right] \leqslant 0 \; \forall k \geqslant \bar{k},$$

which is true by LMI (4.16b). With the assumption on the APRCO attractability which is similar to the QDC attractability in Chap. 2, the feasibility of the APRCO is guaranteed. The proof is complete. ∎

The inequality of APRCO (4.9) will be used by the stabilising agent to derive the stabilisable bounds whenever $\xi_{ic}^k < 0$. The procedure of calculating the stabilisable bounds will, therefore, start from a negative value of ξ_{ic}^k. With the choices of APRCO coefficient matrices according to Theorem 4.1, the existence of a non-empty feasible region is guaranteed. If $\xi_{ic}^k \geqslant 0$ after the time instant $\bar{k}$, the stabilising agent will become inactive. Nevertheless, if $\xi_{ic}^k < 0$ recurs, the stabilising agent will reactivate and override the control moves.

4.4.3 Stabilising Agent with Output-Tracking Algorithm

From the stability condition given in Theorem 4.1, the algorithm of a stabilising agent is outlined in the following procedure:

4.4.4 Control Algorithm

Procedure 4.2 Decentralised Stabilising Agent for Output-Tracking Problems

1. *Off-line:*

   (1) The coefficient matrices Q_{ic}, W_{ic}, S_{ic} for all subsystems $\mathscr{S}_i$ are predetermined from LMIs (4.14) and (4.15).

(2) By choosing $R_{ic}(0) - S_{ic}^T(Q_{ic})^{-1}S_{ic} \succ 0$, the initial condition of $\xi_{ic}^0 > 0$ is deem feasible.

2. *Online:* At every time step $k > 0$,

- *Step* 1- Initiating $k_0(i)$:
  a. Calculate ξ_{ic}^k from the control $\hat{u}_i^{ot}(k)$ generated by the automatic controller or the operator and the known (measurable) output $y_i^{ot}(k)$.
  b. Check $\xi_{ic}^k < 0$, if true, assign k to $k_0(i)$ and go to Step 2, otherwise go back to 1(a).
- *Step* 2- Stabilising bound computation:
  a. Whenever $\xi_{ic}\big(\hat{u}_i^{ot}(k), y_i^{ot}(k)\big) < \gamma_i \times \xi_{ic}\big(u_i^{ot}(k-1), y_i^{ot}(k-1)\big)$,
    i Update Q_{ic} and R_{ic} by solving (4.16). The bias number β_i^k (4.9) is adjusted herein. If $\xi_{ic}^{k-1} \simeq 0 \Rightarrow \beta_i^k \equiv 0$ (using the tuned number μ_i).
    ii Compute the stabilising bound $\tilde{u}_i^{ot}(k)$ by solving the following optimisation:

$$\min_{\tilde{u}_i^{ot}(k)} \|\tilde{u}_i^{ot}(k) - \hat{u}_i^{ot}(k)\|^2,$$
$$\text{s.t. } \xi_{ic}\big(\tilde{u}_i^{ot}(k), y_i^{ot}(k)\big) \geqslant \xi_{ic}\big(\hat{u}_i^{ot}(k-1), y_i^{ot}(k-1)\big) + (\beta_i^k)^2.$$
$$(4.18)$$

    iii Replace $\hat{u}_i^{ot}(k)$ with $\tilde{u}_i^{ot}(k)$ and output to $u_i^{ot}(k)$.
  b. Calculate $\xi_{ic}\big(u_i^{ot}(k), y_i^{ot}(k)\big)$ and memorise the result, then verify

$$\xi_{ic}\big(u_i^{ot}(k), y_i^{ot}(k)\big) \geqslant 0.$$

If true, deactivate the stabilising agent and go back to Step 1(a), otherwise go back to 2(a).

If the trajectory of $u_i^{ot}(k)$ appears chattering around $\bar{u}_i$, the inequality of $\xi_{ic}^k < 0$ can be modified with a tuning number ρ_i to become $\xi_{ic}^k < -\rho_i^2$ in Step 1.b in Procedure 4.2 above.

4.5   Illustrative Examples

4.5.1   Illustrative Example 1—Power Systems

Consider the automatic generation control (AGC) problem in power systems. The electrical frequency of voltage output of a synchronous generator is regulated to maintain a near constant nominal value of 50 or 60 Hz. For a turbine-generation set, a *governor* is installed in a local control panel for this task. The governor is a controller in a feedback control loop having frequency as a controlled variable. For a multiple machine system, which is usually the case in power engineering, such local governors will not be able to eliminate the steady-state offset value, in other words, deviation from the nominal value, when there are non-negligible load changes.

The current technique to solve this problem, i.e. to bring the frequency back to the nominal value within acceptable deviations, is to adjust the output powers of generators at respective control areas. Such adjustments can be computed from the small-signal model (with incremental variables) that also includes the tie-line power flows between the allocated control areas and the mechanical powers required from the corresponding prime-movers—the turbines. This problem is usually referred to as 'Automatic Generation Control', see, e.g. [175], which is a multi-variable *supplement frequency control* problem. It is called 'supplement' because the primary frequency regulation is the responsibility of the local governor.

In this illustrative example, we study the performance of the decentralised model predictive control (DeMPC) with the QDC (without a terminal constraint [93]) for the AGC problem having four control areas and the tie-line connections shown in Fig. 4.8. In this example, the local energy losses can be ignored such that an aggregated model can be used, as shown in Fig. 4.7a, b. The 'small-signal model' for this AGC problem from the power system field is treated as a linear state-space model for the multi-variable control design here, similarly to the work in [167]. The overall (global) open-loop state-space model is unstable with the serial connections.

Distributed model predictive control (DMPC) algorithms have been successfully employed in the AGC problem with tie-lines in the control system literature; see, e.g. [127, 167]. In a DMPC scheme, the communication links among the control areas (subsystems) are often installed and available for data transferring permanently, without any interruptions. In contrast, there are not any communication links between the neighbouring subsystems in a decentralised control scheme. The decentralised

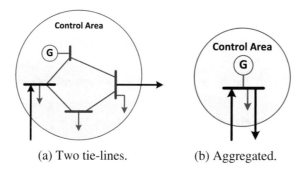

(a) Two tie-lines.   (b) Aggregated.

Fig. 4.7 An example of a control area (subsystem)

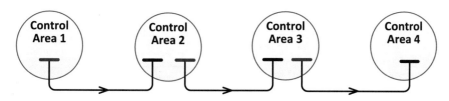

Fig. 4.8 Control areas are serially connected via tie-lines [127, 167]

MPC schemes have not been explored for this particular AGC problem, but for other applications in power systems, such as the one in [95]. Nevertheless, the decentralised MPC has been identified as a potential candidate for power system applications in [57] and elsewhere.

In an AGC application, the electromechanical differential equations for a control area with a single-steam turbine/generator set are given below. This model is extracted from Chap. 10 in [175]. In the state equations, all variables are in small-variation incremental forms. There is only one manipulated variable which is the electrical output power ΔP_{ref_i} required from the local generator. The subscript i represents a local control area i (subsystem i). The state vector of a control area consists of angular frequency $\Delta \omega_i$, mechanical power ΔP_{mech_i}, steam valve position ΔZ_{v_i}, and the incoming tie-line power ΔP_{tie_i} flowing into and out of a control area. There are two coupling variables which are the negative angular frequency $-\Delta \omega_\ell$ and outgoing tie-line power ΔP_{tie_i} from all ν_i connected neighbours.

Nomenclature for the example is as follows:

- ω : Angular frequency of rotating mass
- M : Angular momentum
- D : % in load/ % in frequency
- P_{mech} : Mechanical power
- P_{L} : Non-frequency sensitive load
- T_{ch} : Charging time constant of the prime mover
- Z_{v} : Steam valve position
- P_{ref} : Load reference set point
- T_{G} : Governor time constant
- P_{tie} : Tie-line power flow
- T_{tie} : Tie-line stiffness coefficient, e.g. $T_{\text{tie}} = 377 \times 1/X_{\text{tie}}$ for a 60 Hz system.
- R : Inverse of governor net gain, which affects the change on the output power for a given deviation in frequency.

According to [175], the differential equations of a control area are as follows:

$$\frac{d\Delta \omega_i}{dt} + \frac{1}{M_i}(D_i \Delta \omega_i + \Delta P_{\text{tie}_i} - \Delta P_{\text{mech}_i}) = -\frac{1}{M_i}\Delta P_{\text{L}_i},$$

$$\frac{d\Delta P_{\text{mech}_i}}{dt} + \frac{1}{T_{\text{ch}_i}}\Delta P_{\text{mech}_i} - \frac{1}{T_{\text{ch}_i}}\Delta Z_{v_i} = 0,$$

$$\frac{d\Delta Z_{v_i}}{dt} + \frac{1}{T_{\text{G}_i}}\Delta Z_{v_i} - \frac{1}{T_{\text{G}_i}}\Delta P_{\text{ref}_i} + \frac{1}{R_i T_{\text{G}_i}}\Delta \omega_i = 0,$$

$$\frac{d\Delta P_{\text{tie}_{i+}}}{dt} = \sum_{\ell \in \{\nu_{i_{\text{in}}}\}} T_{\text{tie}_{i\ell}}(\Delta \omega_i - \Delta \omega_\ell),$$

$$\text{for } \nu_i \text{ neighbours, } \nu_i = \nu_{i_{\text{in}}} + \nu_{i_{\text{out}}},$$

$$\Delta P_{\text{tie}_i} = -\Delta P_{\text{tie}_\ell}, \ \Delta P_{\text{tie}_i} = \Delta P_{\text{tie}_{i-}} + \Delta P_{\text{tie}_{i+}}.$$

This is the 'control model' to be used for the control synthesis and design purpose. The control purpose is to minimise the frequency deviation $\Delta\omega_i$ and the tie-line power deviation ΔP_{tie_i} whenever there is a change in the load disturbance ΔP_{L_i} of a control area, ideally back to zero. The manipulated variable ΔP_{ref_i} should deviate from zero as a result, but not be back to zero as in other control problems. The two state variables, mechanical power ΔP_{mech_i} and valve position ΔZ_{v_i}, should also deviate from zero in the occurrence of a permanent load change. It is worth noting that the tie-line power $\Delta P_{\text{tie}_{i_-}}$ can be measured locally, but not the neighbouring frequency $\Delta\omega_{j_-}$. Moreover, transferring accurate frequency measurements across the communication links between power plants in fast real time (within a few seconds), with a high degree of dependability, is among the main difficulties for the power system applications. The decentralised control architecture is, therefore, a preference.

The set point of the MPC in this numerical example is assumed known a priori and has been the steady-state values obtained from the centralised LQR problem, which is as follows:

$$(I - A - B\,K_{LQR})x_{ss} = d, \tag{4.19}$$

where K_{LQR} is the LQR gain for the control law of $u(k) = K_{LQR}x(k)$.

The simulation studies have been conducted with the model parameters and settings as follows: Sampling and updating time: $T_s = 2$. Initial state vectors: $x_1 = 0$; $x_2 = 0$; $x_3 = 0$; $x_4 = 0$. State constraint: $(-\infty, +\infty)$. Control constraint: $|u_i(k)| \leqslant 0.5$. Weighting coefficients: $W_x = \text{diag}\{50\,50\,0\,0\,50\,50\,0\,0\,50\,50\,0\,0\,50\,0\,0\}$, $W_u = \text{diag}\{1\ 1\ 1\ 1\}$. Predictive horizon: $N = 4$. Permanent (constant) load disturbances in this simulation study are as follows: The load in area 2 increases 25% while the load in area 3 decreases 25% from the time Step 2 onward. The weighting coefficients for the two state variables ΔP_{mech} and ΔZ are set to 0 in the simulation studies, similarly to those in [167]. Here, the APRCO coefficient matrices are pre-computed off-line and employed in the stabilising agents for the local MPCs of the DeMPC. The updating steps in Procedure 4.2 are thus not applicable in this example. The trajectories of state and control elements from DeMPC are shown in the following figures.

The centralised MPC with the predictive horizon $N = 12$ provides a desirable performance and is the benchmark for the DeMPC. However, the centralised MPC with $N = 4$ is unstable without any additional constraints as shown in Fig. 4.9. The DeMPC with all local MPC having the predictive horizon $N = 4$ is not unstable, but provides a much worse control performance compared to the centralised MPC with $N = 12$. When the APRCO-based stabilising agent is implemented with Procedure 4.2, the DeMPC is stabilised and further achieves a control performance comparable to that of the centralised MPC with a longer predictive horizon $N = 12$, as shown in Fig. 4.10.

For the evaluation purpose, the APRCO has also been implemented as an enforced stability constraint for the DeMPC similarly to those in Chap. 2. The time responses are given in Fig. 4.11, which also shows a centralised MPC comparable control performance. From these figures, we can see that the APRCO and DeMPC together have successfully delivered a desirable control performance in two implementation

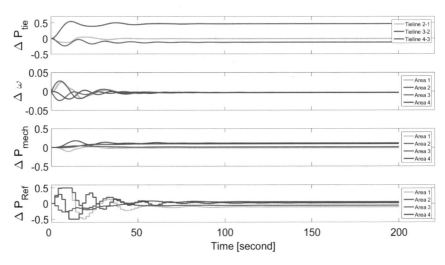

Fig. 4.9 Permanent load disturbance—centralised MPC without any additional constraints, N = 4

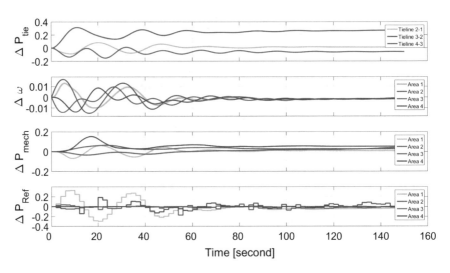

Fig. 4.10 Permanent load disturbance—decentralised MPC with APRCO-based stabilising agent, N = 4. The steady-state values are from the centralised LQR design

cases, stabilising agent and enforced stability constraint. It is worth emphasising here that there are not any exchanged data between the control areas in a DeMPC scheme. There are, however, other partially decentralised control schemes for power systems that minimise the exchanged data between subsystems or control areas, see, e.g. [29, 137], thus require fewer communication links in the cyber-physical system consisting of a power system and a communication network installed for the purpose of real-time control of the power system.

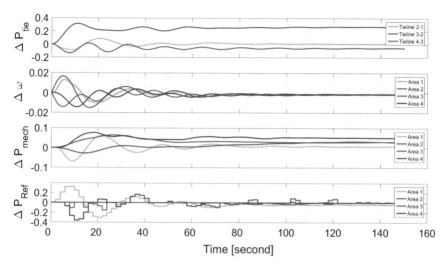

Fig. 4.11 Permanent load disturbance—decentralised MPC with APRCO-based attractivity constraint, N = 4. The steady-state values are from the centralised LQR design

4.5.2 Illustrative Example 2—Network Process System

Consider now the continuous-time state-space models of a typical three-stage countercurrent washing circuit, sketched out in Fig. 4.12 with following configuration matrices:

$$
A_1 = \begin{bmatrix} -29 & 0 & 0 \\ 17 & -6 & 0 \\ -8 & 0 & 0 \end{bmatrix}, \quad
B_1 = \begin{bmatrix} 1 & 0 \\ 0 & 0 \\ 0 & 1 \end{bmatrix}, \quad
E_1 = \begin{bmatrix} 0 & -0.4 \\ 0 & 2 \\ 0 & 0 \end{bmatrix},
$$

$$
A_2 = \begin{bmatrix} -14 & 0 & 0 \\ 18 & -0.5 & 0 \\ -3 & 0 & 0 \end{bmatrix}, \quad
B_2 = \begin{bmatrix} 0 \\ 0 \\ 1 \end{bmatrix}, \quad
E_2 = \begin{bmatrix} 2 & 0 & -1 & 0 \\ 0 & 0 & 0 & 0 \\ 0 & 0.8 & 0 & 0 \end{bmatrix},
$$

$$
A_3 = \begin{bmatrix} -16 & 0 & 0 \\ 12 & -0.7 & 0 \\ -6 & 0 & 0 \end{bmatrix}, \quad
B_3 = \begin{bmatrix} -0.1 & 0 \\ 1 & 0 \\ 0 & 1 \end{bmatrix}, \quad
E_3 = \begin{bmatrix} 2 & 0 \\ 0 & 0.8 \\ 0 & 0 \end{bmatrix},
$$

$$
F_1 = \begin{bmatrix} 0.1 & 0 & 0 \\ 0 & 0 & 1 \end{bmatrix}, \quad
F_2 = \begin{bmatrix} 0.1 & 0 & 0 \\ 0 & 0 & 1 \end{bmatrix}, \quad
F_3 = \begin{bmatrix} 0 & 0.7 & 0 \end{bmatrix},
$$

$$
H = \begin{bmatrix} 0 & 0 & 1 & 0 & 0 \\ 0 & 0 & 0 & 1 & 0 \\ 1 & 0 & 0 & 0 & 1 \\ 0 & 1 & 0 & 0 & 0 \\ 0 & 0 & 1 & 0 & 0 \\ 0 & 0 & 0 & 1 & 0 \end{bmatrix}.
$$

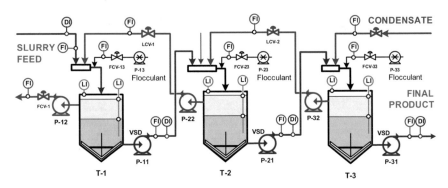

Fig. 4.12 Washing circuit process flow diagram

In this simulation study, the MPC controllers have been set up and coded in MATLAB using the *mpcmove* function. The shorter predictive and control horizons ($N_y = 3$, $N_u = 2$) have been selected for the first two subsystems, while longer predictive horizons ($N_y = 12$, $N_u = 4$) have been selected for the third one. Three subsystems have the same sampling time of 0.4 steps. The weighting coefficients in the *mpcmove* functions are as follows:

$$W_{u_1} = \text{diag}\{2 \quad 1\}, W_{\Delta u_1} = \text{diag}\{0.5 \quad 1\}, W_{y_1} = \text{diag}\{5 \quad 1 \quad 2\},$$
$$W_{u_2} = [1], W_{\Delta u_2} = [0.3], W_{y_2} = \text{diag}\{1 \quad 1 \quad 2\},$$
$$W_{u_3} = \text{diag}\{0.5 \quad 1\}, W_{\Delta u_3} = \text{diag}\{0.2 \quad 0.7\}, W_{y_3} = \text{diag}\{4 \quad 1 \quad 5\}.$$

Figure 4.13a depicts the plant output time responses of three stand-alone subsystems regulated by three separate (stand-alone) controllers applying MPC algorithms. When three tanks are connected into a circuit, the plant output time responses of three interconnection subsystems regulated by the decentralised MPC are shown in Fig. 4.13b showing divergent trajectories. In this simulation, the stabilising agents are not used. To demonstrate the efficacy of Procedure 4.1, we firstly simulate the stabilising agent using only off-line-computed coefficient matrices for the quadratic constraint (i.e. the coefficient matrices of the quadratic constraint are not updated online as required by Step 2 of Procedure 4.1).

A simple vertical and horizontal line search algorithm has been implemented to find a feasible solution to the problem (4.1) in this simulation. When the MPCs of subsystems are not imposed with stability constraints (Step 3 of Procedure 4.1 is ignored), the initial time $k_0(k)$ of subsystem 1 appears at steps 11, 58, 73, 105 and 185. The time instant $\bar{k}(k)$ of subsystem 1 incurs at the time steps 28, 70, 84 and 175. The initial time $k_0(k)$ of subsystem 3 occurs at the time steps 10, 30 and 71, and the time instant $\bar{k}(k)$ of subsystem 3 takes place at the time steps 19 and 67. This means the quadratic constraint in this example is the REDC defined in Chap. 2 (Fig. 4.14).

When the coefficient matrices of the quadratic constraint are updated online (i.e. the full Procedure 4.1 is employed), the plant output and control time responses are shown in Fig. 4.15, also indicating attractive trajectories. Moreover, they exhibit

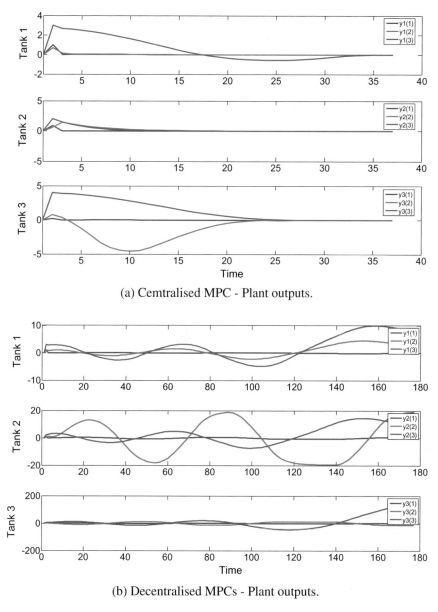

(a) Cemtralised MPC - Plant outputs.

(b) Decentralised MPCs - Plant outputs.

Fig. 4.13 Decentralised MPCs without stabilising agents

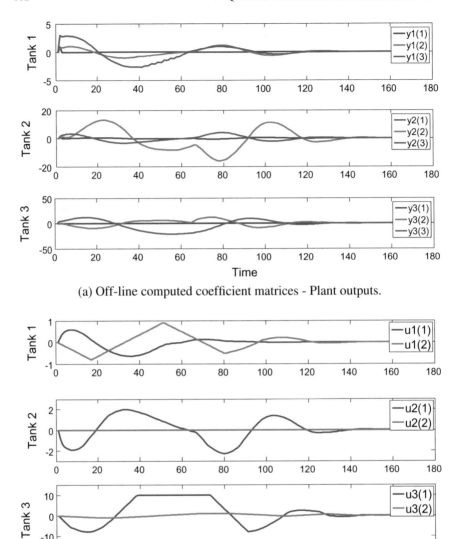

(a) Off-line computed coefficient matrices - Plant outputs.

(b) Off-line computed coefficient matrices - Control inputs.

Fig. 4.14 Decentralised stabilising agents with off-line-computed coefficient matrices

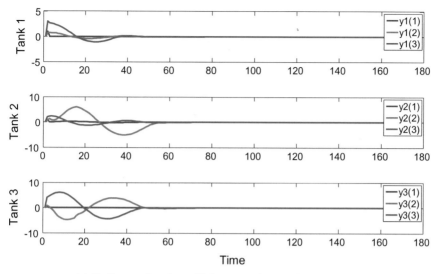

(a) Online updated coefficient matrices - Plant outputs.

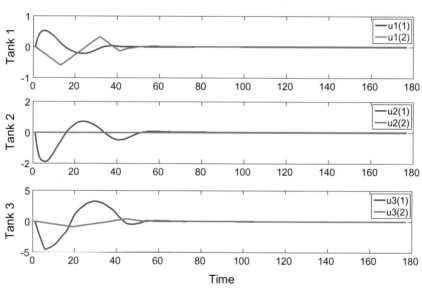

(b) Online updated coefficient matrices - Control inputs.

Fig. 4.15 Decentralised stabilising agents with online updated coefficient matrices

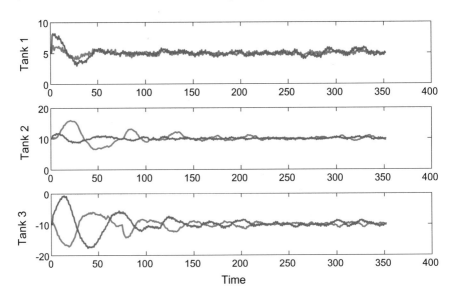

Fig. 4.16 Decentralised stabilising agents with a nonzero output set point and input disturbances

a significant performance improvement compared to those from the off-line solution shown in Fig. 4.13. There are not any wind-up actions in these control profiles
when the coefficient matrices of the quadratic constraint are updated online as in
Procedure 4.2. The nonzero set point case is demonstrated in Fig. 4.16 with state
disturbances.

This numerical example has shown that the *stabilising agent* approach to the
stabilisation of interconnected systems with decentralised MPC is computationally
effective. Notably, the control performance with the online updated coefficient matrices for the quadratic constraint is better than those from the off-line solution in this
second example. Both the settling time and the state peaks in Fig. 4.15 are reduced
relatively to those in Fig. 4.13.

4.6   Concluding Remarks

In this chapter, a constructive approach of stabilising agents has been developed from
the quadratic constraint (APRC and QDC) for the feasible implementation. The stabilising bound on the immediate future value of the control vector is updated on
the fly in this approach. The quadratic constraint is not converted into an integrated
constraint for the MPC optimisation problem, but implemented in a segregated manner. The stabilising agent that operates independently to, and separated from, the
control algorithm will override the control actions if they are determined as beyond a
tuned stabilising boundary of the quadratic constraint, either APRC or QDC. Being

disparate from the associated controller, the stabilising agent is inter-operable with different control algorithms.

The method of stabilising a controlled system with the stabilising agent has been inspired by the segregation principle in the design of reliable computer systems and networks. The segregation of a stabilising agent from its associated control system facilitates a semi-automatic operational system that accommodates both automatic and human-in-the-loop controllers.

This chapter also presents the development with the steady-state-independent dissipative criteria and quadratic constraint for the output-tracking control problem of interconnected systems subject to input disturbances. The two numerical examples have demonstrated that the stabilising agent approach can be successfully deployed for interconnected systems together with a decentralised MPC scheme.

Chapter 5
Quadratic Constraint with Data Losses

5.1 Introduction

The quadratic constraints, APRC or QDC, presented in Chap. 2 are extended in this chapter for networked control systems (NCS) and applied to a partially decentralised model predictive control scheme for interconnected systems having multiple communication network topologies. The decentralised MPC approach has been employed in previous chapters. This chapter introduces the data from neighbouring subsystems to the quadratic constraint to improve the control performance. These information data are available via the communication channels among the subsystems. The intermittent data losses are also inclusive in the problem formulation in this chapter. The conceptual system architecture shown in Fig. 5.1 is considered in the development. There are three networks in this block diagram:

1. The subsystem 'network' (dynamic couplings) of the interconnected system as has been considered in previous chapters;
2. The sensor and device network (the middle block) between the local subsystems and the corresponding controllers;
3. And the communication network between the controllers (the top block) that has multiple connection topologies.

The measurement data and control signals are transferred between the process plants and the controllers via the sensor and device network represented by the block above the subsystems in Fig. 5.1. The cooperative data among subsystems are also exchanged among the neighbouring subsystems via the controller communication network represented by the block on the top of MPC controllers in Fig. 5.1. In this development, the intermittent data losses may incur to both sensor and device network in the middle and in the controller communication network on the top. Further, the controller communication network may be broken down in long durations.

For guaranteeing the system attractability in a decentralised architecture, the stability condition is established from the *'data-lost robust dissipativity'* of the

© Springer Nature Singapore Pte Ltd. 2018
A. Tri Tran C. and Q. Ha, *A Quadratic Constraint Approach to Model Predictive Control of Interconnected Systems*, Studies in Systems, Decision and Control 148, https://doi.org/10.1007/978-981-10-8409-6_5

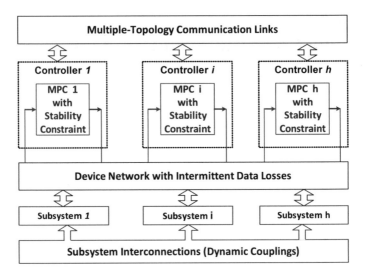

Fig. 5.1 Conceptual block diagram of the system architecture

controlled subsystems and the satisfaction of decoupled quadratic constraints. The intermittent losses of state data and control signals are effectively compensated for right after the dropout occurrences. The last received value of a control vector is used during the interruption intervals. Alternatively, if the communication traffic and local memory allow for retaining the planned control moves, then the open-loop predictive result from the MPC optimisation can be used. The quadratic constraint is characterised such that the stabilising bound is assuredly feasible in the presence of data losses.

The stability of feedback control systems operating in an imperfect shared communication network has been attracting much research since the problem was raised in [16, 55]. In such a networking environment, the output measurements from transmitters as well as the control signals to actuators are subject to occasional delays and random data losses because of the communication data dropouts, limited bandwidths that may cause congestions, and transmission errors. For example, the checksum bits in the data-link layer of an IP-based protocol are checked by the receiver for non-error data packets. If parity is not met, the packet will be dropped. The performance of an NCS is therefore easily degraded, and in some critical scenarios, the stability may not be guaranteed.

By modelling the networked control systems with data losses and transmission delays as 'asynchronous multi-rate sampled-data systems' whose controllers do not receive the measured data from sensors at fixed time instants, the authors in [56] have referenced the problem back to previous research work in the early 1980s. Since then, the analysis and design of NCS using state feedback and output feedback methods to deal with either data losses and random delays, or both, in various optimal control schemes have been developed and presented, such as those in [99, 170, 174]. These approaches ensure that the closed-loop system stability is maintained if the network frozen or inaccessible time period is shorter than the maximum allowable transmission interval (MATI). Both deterministic and stochastic models have already been used for analysing and synthesising the NCSs. In deterministic approaches, the packet drop and asynchronously sampling processes can be adequately modelled to formulate a robust control problem [170]. The stochastic control approaches to NCSs have been presented recently in [28, 84, 177]. The research in NCSs is quite mature in its own right with recent results published in, e.g. [27, 50, 58, 61, 87, 98, 108, 132, 149, 150, 156, 169, 171, 181] and references thereof. The NCS is apparently a current research theme in the field of systems and control.

Apart from the classical closed-form methods, the model predictive control strategy has been recognised as suitable for the data loss problem when the future control moves can be memorised using the smart actuators, see, e.g. [148]. The minimum requirement for the closed-loop system stability remains always that the control horizon is not less than MATI. Instead of sending the whole control sequence, the last received values were used in [107]. Different MPC strategies for the robust stability, quantisation errors as well as for reducing network traffics in linear systems have been presented in [43, 51, 87] and others. In [35, 83, 119], the stability guaranteed methods for NMPC have been extended to NCS. The constraints that define the optimisation problems and the implementation procedures have been modified to account for data losses in these approaches. The prediction inconsistency or trajectory drifts between the old and new predictions has been analysed in [35] to maintain the NMPC stability. The input-to-state stability conditions for NMPC have been developed with and without the time-stamping information in [113, 119].

While much research has been focused on the data packet dropouts and transmission delays, the problem of assuring the stability for interconnected systems with communication disconnections between subsystems has not been addressed up until recently [78]. In [126], the graph theory applying to the coordination controls of autonomous vehicles in an imperfect data environment has been discussed. The approach of dynamic graphs for interconnected systems with different coupling structures is introduced in [141]. The cooperative control schemes for multi-agent systems with time-varying communication topologies have been developed in [33, 34, 69, 104]. These approaches do not, nonetheless, accommodate the transmission packet dropouts in local sensor networks.

The developments in this chapter address the flexible communication links between subsystems and the data losses in sensor and actuator networks. The system and network models are outlined in Sect. 5.2. Section 5.3 considers only local data losses, wherein the data-lost robust dissipative criteria are derived for use with the quadratic constraint. The stability conditions are then presented in Sect. 5.4. Both issues of local data losses and flexible communication links are resolved in Sect. 5.5 by a perturbed cooperative-state feedback strategy for the partially decentralised MPC scheme.

5.2   System and Networked Control Models

The system models in Chap. 2 are reused in this chapter. However, they are presented with new reference numbers in this section, for completeness.

5.2.1   System Model

The interaction-oriented model [86] is employed to describe the input and output nature of interconnected systems. Consider an interconnected system Σ consisting of h subsystems, each denoted as $\mathscr{S}_i$, $i = 1, \ldots, h$ and has a discrete-time state-space model of the form:

$$
\mathscr{S}_i : \begin{cases} x_i(k+1) = A_i x_i(k) + B_i u_i(k) + E_i v_i(k), \\ y_i(k) = C_i x_i(k), \quad w_i(k) = F_i x_i(k), \end{cases} \tag{5.1}
$$

where $u_i \in \mathbb{U}_i \subset \mathbb{R}^{m_i}$, $y_i \in \mathbb{R}^{q_i}$, $x_i \in \mathbb{R}^{n_i}$, $v_i \in \mathbb{R}^{m_{v_i}}$, $w_i \in \mathbb{R}^{q_{w_i}}$; x_i, u_i, y_i, v_i and w_i are, respectively, state, control, output, interaction input and interaction output vectors. In this chapter, the following ball constraint is considered

$$
\mathbb{U}_i = \{u_i : \|u_i\|^2 \leqslant \eta_i\}, \ \eta_i > 0. \tag{5.2}
$$

When $v_i(k) = w_j(k)$, $j \neq i$, we say that the subsystems $\mathscr{S}_i$ and $\mathscr{S}_j$ are linearly connected to each other. The subsystem connections are encapsulated by the global interconnection matrix H, also called global coupling matrix. The elements of H are either 1 or 0 only, with 1 representing a subsystem connection.

The larger-scale global system Σ is represented by a state-space model of the block diagonal system $\mathscr{S}$ formed by h subsystems $\mathscr{S}_i$ and their interconnection matrix H, as follows:

$$
\Sigma : \begin{cases} x(k+1) = A x(k) + B u(k) + E v(k), \\ y(k) = C x(k), \quad w(k) = F x(k), \\ v(k) = H w(k), \end{cases} \tag{5.3}
$$

where $A = \mathrm{diag}[A_i]_1^h$, $x = [x_1^T \ldots x_h^T]^T$, and similarly for other block diagonal matrices B, C, E, F and stacking vectors u, y, v, w of Σ. From a centralised point of view, the system Σ can be rewritten as

$$
\Sigma : \begin{cases} x(k+1) = (A + EHF) x(k) + B u(k) = A_\Sigma x(k) + B u(k), \\ y(k) = C x(k). \end{cases} \tag{5.4}
$$

Fig. 5.2 Networked control system—a single loop

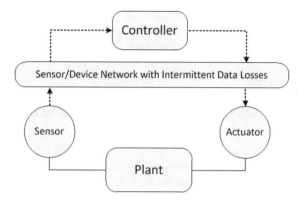

In a decentralised control scheme, the model of each stand-alone subsystem is requisite. When all the interactive inputs and outputs vanish (i.e. $v_i = 0$ and $w_i = 0$), a stand-alone subsystem $\mathscr{S}_i$ has the state-space model of the form:

$$\mathscr{S}_i|_{\text{stand alone}} : \begin{cases} x_i(k+1) & = A_i x_i(k) + B_i u_i(k), \\ y_i(k) & = C_i x_i(k). \end{cases} \tag{5.5}$$

The following properties are assumed in this development:

1. (A, C) and (A_i, C_i) are observable, while (A, B) and (A_i, B_i) are controllable.
2. The updating instants are synchronised between all subsystems and their controllers.

5.2.2 Deterministic Data-Lost Process

In this chapter, a sensor/device network is installed for the global system for updating the measurements and actuating the control devices, as illustrated in Fig. 5.2. The sampling and updating time instants are assumed synchronised between all controllers. The output vector y of Σ becomes $\tilde{y}$ beyond the network interface ports inside the controllers. It is assumed that the data losses are identical to all variables by hardware and firmware configurations. When the output vector $y(k)$ at the time instant k is transmitted successfully, we have $\tilde{y}(k) = y(k)$, otherwise $\tilde{y}(k) = y(k-1)$ (i.e. when the data are lost).

For the manipulated variables, the last received values are applied to the plant (e.g. by industrial smart actuators) during the data loss intervals. By representing the consecutive updating instants of $\tilde{y}(k)$ with a sequence of integer numbers $\mathscr{J} = \{j_1, \ldots, j_q, \ldots, j_\kappa, \ldots\}$, $\mathscr{J} \subset \mathbb{Z}^+$, the time interval between j_q and j_{q+1} can be treated as one transmission period. If the data are received at the time step k, we have $j_\kappa = k$. The upper bound of the successful transmission intervals is denoted as μ

(or MATI—maximum allowable transmission interval).

$$\mu := \max_{j_q \in \mathscr{J}} \left(\tau(q) \right), \quad \tau(q) := j_q - j_{q-1}. \tag{5.6}$$

5.3   Dissipative Condition for Networked Control Systems

The dissipation inequality and the affirmatively APRC of each subsystem $\mathscr{S}_i^s$ (5.1) are presented in this section. The global system Σ (5.4) is said to be *data-lost robustly quadratically dissipative* with respect to the quadratic supply rate $\xi(y(k), u(k))$ defined as,

$$\xi(y(k), u(k)) := [y(k)^T u(k)^T] \begin{bmatrix} Q & S \\ S^T & R \end{bmatrix} \begin{bmatrix} y(k) \\ u(k) \end{bmatrix}, \tag{5.7}$$

where Q, R, S are multiplier matrices with symmetric Q and R, if there exists a non-negative C^1 function, addressed as storage function, $V(x(k))$, $V(0) = 0$, such that for all $u(k)$ and all $k \in \mathbb{Z}^+$, the following dissipation inequality is satisfied irrespectively of the initial value of the state $x(0)$:

$$V\left(x(k+\tau)\right) - \alpha \times V\left(x(k)\right) \leqslant \xi\left(y(k), u(k)\right), \\ \forall \tau \in \{1, 2, \ldots, \mu\}, \quad 0 < \alpha < 1. \tag{5.8}$$

The square storage function of the form $V\left(x(k)\right) = x(k)^T P x(k)$, where $P = P^T \succ 0$, is considered in this chapter. Here, define another quadratic function ξ_{ic}, addressed as supply rate of the controlled subsystem $\mathscr{S}_i$, as

$$\xi_{ic}\left(u_i(k), y_i(k)\right) := [u_i(k)^T y_i(k)^T] \begin{bmatrix} Q_{ic} & S_{ic} \\ * & R_{ic}(k) \end{bmatrix} \begin{bmatrix} u_i(k) \\ y_i(k) \end{bmatrix}. \tag{5.9}$$

For conciseness, $\xi_{ic}\left(u_i(k), y_i(k)\right)$ is denoted as ξ_{ic}^k; $\xi\left(y(k), u(k)\right)$ is denoted as ξ^k; and the time index k and j_q (q is the index of the data-lost process of MATI=μ) are represented by the superscript typefaces k and q, respectively, without brackets where appropriated.

Definition 5.1 The controlled $\mathscr{S}_i$ (5.1) is said to satisfy the affirmatively APRC with respect to $\xi_{ic}(u^q, y^q)$ (5.9), or simply aAPRC, in the presence of data-lost process of MATI=μ, if there is $j_{q0} \in \mathscr{J}, 0 \leqslant \gamma_i^q < 1$ and $|\beta_i^q| \geq \sqrt{(1 - \gamma_i) \times |\xi_i^{q-1}|}$, such that

$$\begin{cases} \xi_{ic}^q \geqslant 0, & \text{if } \xi_{ic}^{q-1} \geqslant 0, \\ \xi_{ic}^q \geqslant \xi_{ic}^{q-1} + (\beta_i^q)^2, & \text{if } \xi_{ic}^{q-1} < 0, \end{cases} \quad \forall j_q > j_{q0}. \tag{5.10}$$

Remark 5.1 Equation (5.10) can also be expressed as follows:

$$\xi_{ic}^q \geqslant \frac{1 - \mathrm{sgn}(\xi_{ic}^{q-1})}{2} \times \left(\xi_{ic}^{q-1} - \mathrm{sgn}(\xi_{ic}^{q-1}) \times (\beta_i^q)^2\right) \forall j_q > j_{q0}. \qquad (5.11)$$

The bias number β_i^q may be different at every successful transmission instants j_q. The condition of aAPRC is thus dynamically established as there may be several time instants j_{q_0} during the lifetime of a control subsystem at which $\xi_{ic}^{q_0} < 0$, similarly to the REDC in Chap. 2.

5.4 Stability Condition for Networked Control Systems

By denoting $Q_c := \mathrm{diag}[Q_{ic}]_1^h$, $S_c := \mathrm{diag}[S_{ic}]_1^h$, $R_c := \mathrm{diag}[R_{ic}]_1^h$, and

$$\xi_c(y, u) := u^T Q_c u + 2y^T S_c u + y^T R_c y,$$

as the supply rate of Σ, $u \in \mathbb{R}^m$, the aAPRC-based attractivity condition for Σ subject to the data-lost process (5.6) is stated in the next theorem.

Denote $\varphi_i^q := \frac{1}{2}(1 - \mathrm{sgn}(\xi_{ic}^q)) \times \xi_{ic}^q$, $\theta_i^q := \varphi_i^q \times \mathrm{sgn}(\xi_{ic}^{q-1}) \times (\beta_i^q)^2$.

Theorem 5.1 *Consider the system Σ (5.4) and the data-lost process (5.6). Suppose that*

(i) P, Q, S, R solve the following LMIs:

$$\begin{bmatrix} A_\Sigma^{\tau T} P A_\Sigma^\tau - \alpha \times P - C^T Q C & A_\Sigma^{\tau T} P \mathscr{B} - C^T S M \\ * & \mathscr{B}^T P \mathscr{B} - M^T R M \end{bmatrix} \prec 0, \ P \succ 0, \qquad (5.12)$$

$$\forall \tau \in \{1, 2, \ldots, \mu\},$$

where $M = [0 \ \ldots \ 0 \ I_m]$ and $\mathscr{B} = [A_\Sigma^{\tau-1} B_\Sigma \ A_\Sigma^{\tau-2} B_\Sigma \ \ldots \ A_\Sigma B_\Sigma \ B_\Sigma]$;

(ii) Whenever $\varphi_i^{q-1} < 0$, Q_{ic}^q and R_{ic}^q solve the following LMIs:

$$\begin{bmatrix} y_i^{q T}(Q_i + R_{ic}^q)y_i^q & y_i^{q T} S_i \\ S_i^T y_i^k & R_i + Q_{ic}^q \end{bmatrix} \prec 0, \ Q_{ic}^q \prec 0, \qquad (5.13a)$$

$$y_i^{q T} R_{ic}^q y_i^q - \varphi_i^{q-1} + \theta_i^q \geqslant 0; \qquad (5.13b)$$

and S_{ic}^q is chosen as a null space of y_i^T, i.e. $S_{ic}^q y_i^T = 0$;

(*iii*) *There exists a compact neighbourhood of the origin* $\mathbb{X}$ *such that the controlled system* (5.4) *is aAPRC-attractability in* $\mathbb{X}$ *with some h control sequences* $\{u_i(j_q) \in \mathbb{R}^{m_i}\}$, $i = 1, 2, \ldots, h$, *as defined in Chap. 2.*

Then, any aAPRC-attractable control sequences $\{u_i(j_q) \in \mathbb{R}^{m_i}\}$, $i = 1, 2, \ldots, h$, *satisfying the respective aAPRC* (5.10), *stabilise the unconstrained* Σ *subject to the data-lost process* (5.6), *in the sense that* $\|x\| \to 0$ *as* $k \to \infty$.

Proof (i) *Dissipativity*: The global system Σ is data-lost robust dissipative if the LMI (5.12) is fulfilled. This LMI is derived from the dissipation inequality (5.8) using the predictive state vector of the form $x(j + \tau + 1) = A_{\Sigma}^{\tau} x(j) + \mathscr{B}\breve{u}(j)$, where $\breve{u}(j) := [u(j)^T u(j + 1)^T \ldots u(j + \tau)^T]^T$.

(ii) *Stability*: The proof starts at some time instant when $\xi_c^q < 0$. It is apparently that ξ_c^q will eventually go to the vicinity of zero and become non-negative with $|\beta_i^q| \geqslant \sqrt{(1 - \gamma_i)} \times |\xi_i^{q-1}|$, as the second inequalities of aAPRC in (5.10) of h subsystems are feasible for all $q > q_0$ which are given by LMIs (5.16) and proved in (iii) below. Similarly to Theorem 5.1,

$$V(x^{q+1}) - \alpha \times V(x^q) \leqslant (*) \begin{bmatrix} y^{q T}(Q + R_c^q) y^q & y^{q T}(S + S_c) \\ * & R + Q_c \end{bmatrix} \begin{bmatrix} I \\ u^q \end{bmatrix} + \max(\gamma_i) \times \varphi(j_q),$$

where $V(x^q) := x^{q T} P x^q$.

(iii) *Feasibility for aAPRC*: (5.13) is a direct result of the first control qualified condition in Chap. 2 ∎

In the following, the constraint on the control increment $\Delta u_i(k)$, $\Delta u_i(k) := u_i(k) - u_i(k - 1)$, is taken into account in the attractivity condition. Consider the bounded constraint below,

$$\mathbb{U}_{\Delta i} := \{\Delta u_i : \|\Delta u_i\|^2 \leqslant \eta_{\Delta i}\}. \tag{5.14}$$

The attractivity condition for Σ subject to the data-lost process (5.6) and two constraints (5.2) and (5.14) is stated in the next theorem.

Theorem 5.2 *Consider* Σ *(5.4) and the data-lost process (5.6). Suppose that*

(i) *P, Q, S, R solve the following LMIs:*

$$\begin{bmatrix} A_{\Sigma}^{\tau T} P A_{\Sigma}^{\tau} - \alpha P - C^T Q C & A_{\Sigma}^{\tau T} P \mathscr{B} - C^T S M \\ * & \mathscr{B}^T P \mathscr{B} - M^T R M \end{bmatrix} \prec 0, \ P \succ 0, \tag{5.15}$$

$$\forall \tau \in \{1, 2, \ldots, \mu\};$$

(ii) *Whenever* $\varphi_i^{q-1} < 0$, Q_{ic}^q, S_{ic}^q, R_{ic}^q *and* α_{di}^q *solve the following online LMIs:*

$$\begin{bmatrix} y_i^{q^T}(Q_i + R_{ic}^q)y_q^k & y_i^{q^T}(S_i + S_{ic}^q) \\ * & R_i + Q_{ic}^q \end{bmatrix} \prec 0, \quad Q_{ic}^q \prec 0, \tag{5.16a}$$

$$y_i^{q^T} R_{ic}^q y_i^q - \varphi_i^{q-1} - \theta_i^q \geqslant 0; \tag{5.16b}$$

$$\begin{bmatrix} -\alpha_{d_i}^q \times \eta_{\Delta i} - \varphi_i^{q-1} + \theta_i^q + (*)\mathscr{R}_{di} \begin{bmatrix} x_i^{q-1} \\ u_i^{q-1} \end{bmatrix} & \begin{bmatrix} u_i^{(q-1)T}(B_{di}^T S_{ic}^q + M_\tau^T Q_{ic}^q) \\ +x_i^{(q-1)T} A_{di}^T S_{ic}^q \end{bmatrix} \\ * & Q_{ic}^q + \alpha_{d_i}^q I \end{bmatrix} \succ 0,$$

$$\alpha_{d_i}^q > 0, \tag{5.16c}$$

where

$$\mathscr{R}_{di} := \begin{bmatrix} A_{di}^T R_{ic}^q A_{di} & A_{di}^T R_{ic}^q B_{di} + A_{di}^T S_{ic}^q M_\tau \\ * & B_{di}^T R_{ic}^q B_{di} + B_{di}^T S_{ic}^q M_\tau + M_\tau^T Q_{ic}^q M_\tau \end{bmatrix},$$

$$A_{di} = C_i A_i^\mu, \quad B_{di} = C_i \mathscr{B}_{di},$$

$$\mathscr{B}_{di} := [A_i^{\mu-1} B_{di} \ A_i^{\mu-2} B_{di} \ \ldots \ A_i B_{di} \ B_{di}],$$

$$B_{di} := [B_i \ E_i], \quad M_\tau = [0_{m_i} \ \ldots \ 0_{m_i} \ I_{m_i} \ 0_{m_{v_i}} \ \ldots \ 0_{m_{v_i}}],$$

$$u_i^{q-1} = M_\tau \breve{u}_{di}, \quad \breve{u}_{di} := [\breve{u}_i \ \breve{v}_i];$$

(*iii*) *There exists a compact neighbourhood of the origin* $\mathbb{X}$ *such that the controlled system (5.4) is aAPRC-attractability in* $\mathbb{X}$ *with some h control sequences* $\{u_i(j_q) \in \mathbb{R}^{m_i}\}$, $i = 1, 2, \ldots, h$, *as defined in Chap. 2.*

Then, any aAPRC-attractable control sequences $\{u_i(j_q) \in \mathbb{U}_i | \ \Delta u_i(j_q) \in \mathbb{U}_{\Delta_i}\}$, $i = 1, 2, \ldots, h$, *satisfying the respective aAPRC (5.10), stabilise* Σ *subject to the data-lost process* μ *(5.6), in the sense that x remains finite and* $\|x\| \to 0$ *as* $k \to \infty$.

Proof The proof of this theorem is identical to that of Theorem 5.1, except for the last LMI (5.16c) which is derived as follows:

$$\xi_{ic}^q \geqslant \sigma_i^q \times \xi_{ic}^{q-1} - \theta_i^q$$

$$\Leftrightarrow (x_i^{(q-1)T} A_i^{\tau T} + \breve{u}_{di}^{(q-1)T} \mathscr{B}_{di}^T)C_i^T R_{ic}^q C_i(A_i^\tau x_i^{q-1} + \mathscr{B}_{di}\breve{u}_{di}^{q-1})$$

$$+ (x_i^{(q-1)T} A_i^{\tau T} + \breve{u}_{di}^{(q-1)T} \mathscr{B}_{di}^T)C_i^T S_{ic}^q(M_\tau \breve{u}_{di}^{q-1} + \Delta_{di}^q)$$

$$+ (\breve{u}_{di}^{(q-1)T} M_\tau^T + \Delta u_i^{q^T})Q_{ic}^q(M_\tau \breve{u}_{di}^{q-1} + \Delta u_{di}^q) \geqslant \sigma_i^q \times \xi_{ic}^{q-1} + \sigma_i^q \times (\beta_i^q)^2,$$

$$\Leftrightarrow \quad x_i^{(q-1)T}(A_{di}^T R_{ic}^q A_{di})x_i^{q-1} + \breve{u}_{di}^{(q-1)T}(B_{di}^T R_{ic}^q B_{di} + B_{di}^T S_{ic}^q M_\tau + M_\tau^T Q_{ic}^q M_\tau)\breve{u}_{di}^{(q-1)}$$

$$+ 2x_i^{(q-1)T}(A_{di}^T R_{ic}^q B_{di} + A_{di}^T S_{ic}^q M_\tau)\breve{u}_{di}^{(q-1)} + \Delta u_i^{qT} Q_{ic}^q \Delta u_i^{qT}$$

$$+ 2x_i^{(q-1)T}(A_{di}^T S_{ic}^q)\Delta u_i^{qT} + 2\breve{u}_i^{(q-1)T}(B_{di}^T S_{ic}^q + M_\tau^T Q_{ic}^q)\Delta u_i^{qT}$$

$$\geqslant \sigma_i^q \times \xi_{ic}^{q-1} - \theta_i^q, \text{ where } A_{di} = C_i A_i^\mu, \ B_{di} = C_i \mathscr{B}_{di}$$

$$\Leftrightarrow \quad (*)Q_{ic}^q \Delta u_i^q + 2[\breve{u}_i^{(q-1)T}(B_{di}^T S_{ic}^q + M_\tau^T Q_{ic}^q) + x_i^{(q-1)T}(A_i^T C_i^T S_{ic}^q)]\Delta u_i^q$$

$$+ \begin{bmatrix} x_i^{q-1} \\ \breve{u}_i^{(q-1)} \end{bmatrix}^T \begin{bmatrix} A_{di}^T R_{ic}^q A_{di} & A_{di}^T R_{ic}^q B_{di} + A_{di}^T S_{ic}^q M_\tau \\ * & B_{di}^T R_{ic}^q B_{di} + B_{di}^T S_{ic}^q M_\tau + M_\tau^T Q_{ic}^q M_\tau \end{bmatrix} \begin{bmatrix} x_i^{q-1} \\ \breve{u}_i^{(q-1)} \end{bmatrix}$$

$$- \sigma_i^q \times \xi_{ic}^{q-1} + \theta_i^q \geqslant 0.$$

The last inequality holds true with every $\Delta u_i^q \in \mathbb{U}_{\Delta_i}$ (5.14) by the feasibility of the LMI (5.16c). ∎

The decentralised model predictive control problem with the quadratic constraint at every time step is formulated similarly to those in previous chapters, in which the coefficient matrices of the supply rate are determined from the attractability conditions in this section.

A cooperative strategy for the networked control interconnected systems is outlined in the next section.

5.5 Perturbed Cooperative-State Feedback Strategy for Interconnected Systems

A perturbed cooperative-state feedback (PSF) strategy is presented in this section for the control of interconnected systems that have multiple connection topologies. In this approach, the demands on communication links and online data are limited with the PSF which consists of a cooperative-state feedback and a perturbation variable, e.g. $u_i = K_{ij}x_j + w_i$, whose state feedback gains are determined in accordance with the network topologies while the perturbation is computed by the online optimisation adopting only decoupled objective functions [157]. The PSF strategy can resolve both issues, the sensor data losses and the communication network breaks, thanks to the two components of the control. For the implementation, a partially decoupled attractivity constraint for the decentralised model predictive controllers is subsequently developed from the underlying PSF strategy that engages off-line and online computations. Numerical simulation for the automatic generation control problem is studied at the end of this chapter to illustrate the effectiveness of the presented PSF strategy in a realistic engineering problem.

Control applications in the imperfect data environments are flourishing with the Internet-based control and wireless networking systems. The control design and architectures for interconnected systems whose subsystems are dynamically coupled are thus associated with a communication network in many modern applications. The interconnected systems such as power systems or large chemical processes that operate in harmony with the sensor and subsystem communication networks can be viewed as parts of a cyber-physical system, as defined by the National Institute of Standards and Technology (NIST)—'a co-engineered interacting network of physical and computational components'.

In a fully *decentralised* control scheme, the communication links between subsystems are totally disconnected [131] which may affect the overall control performance for strongly coupled subsystems. On the other hand, as the communication links are required to be in place continuously, a communication interruption may lead to the closed-loop system instability in a *distributed* control scheme. However, both the total disconnections and permanently connected links are seldom the reality scenarios. The subsystem communication network is neither persistently uninterrupted nor available with a never-changing network topology in practice.

The PSF strategy is introduced in this chapter as an alternative approach to the existing distributed strategies that require such an uninterrupted communication network and system. The presenting strategy only requires the data to be exchanged between neighbouring subsystems whenever the corresponding communication channels are available, hence, leverage the disadvantages of a fully decentralised approach and the advantages of a cooperative system. The demand for data exchanged between subsystems is reduced with the PSF strategy while the available cooperative data are still capitalised on, as a consequent. The PSF strategy is then applied to a partially decentralised model predictive control scheme for the cyber-physical and interconnected systems.

5.5.1 Subsystem Model

Consider an interconnected system Σ consisting of h nodes representing h subsystems, each denoted as $\mathscr{S}_i$, indexed by the elements of the set $\mathscr{N} := \{1, \ldots, h\}$, and has a discrete-time state-space model of the form:

$$\mathscr{S}_i : \quad x_i(k+1) = A_i x_i(k) + B_i u_i(k) + E_i v_i(k), \tag{5.17}$$

where $A_i \in \mathbb{R}^{n_i \times n_i}$, $B_i \in \mathbb{R}^{n_i \times m_i}$, $E_i \in \mathbb{R}^{n_i \times m_{v_i}}$; $u_i \in \mathbb{R}^{m_i}$ and $x_i \in \mathbb{R}^{n_i}$ are local control and state vectors, respectively; and $v_i \in \mathbb{R}^{m_{v_i}}$ is the interactive, or dynamic coupling, vector.

5.5.2 Network Graph

A network system is identified by a graph $G = (\mathscr{N}, \mathscr{S}, \mathscr{E})$, where $\mathscr{S} = \{\mathscr{S}_1, \ldots, \mathscr{S}_n\}$ is the set of nodes, and $\mathscr{E} \subseteq \{(\mathscr{S}_i, \mathscr{S}_j) : \mathscr{S}_i, \mathscr{S}_j \in \mathscr{S}, i, j \in$

$\mathcal{N} \mid i \neq j\}$ is the set of edges. The graph is undirected (or bidirectional) if for any $i, j \in \mathcal{N}$, $(\mathcal{S}_i, \mathcal{S}_j) \in \mathcal{E} \Leftrightarrow (\mathcal{S}_j, \mathcal{S}_i) \in \mathcal{E}$. A node $\mathcal{S}_i \in \mathcal{N}$ is connected to a node $\mathcal{S}_j \in \mathcal{N}$ if there is a path from $\mathcal{S}_i$ to $\mathcal{S}_j$ in the graph following the orientation of the arcs.

Neighbours: If $(\mathcal{S}_i, \mathcal{S}_j) \in \mathcal{E}$, we say that $\mathcal{S}_j$ is neighbour to $\mathcal{S}_i$. The set of all neighbours to node $i \in \mathcal{N}$ is denoted as $\mathcal{N}_i(G) = \{j, j \in \mathcal{N} : (\mathcal{S}_i, \mathcal{S}_j) \in \mathcal{E}\}$. Herein, we consider a network of fixed subsystem dynamic couplings, or inter-connections, as shown in Fig. 5.1, denoted as $G_s = (\mathcal{N}, \mathcal{S}, \mathcal{E}_s)$, and a flexible inter-system communication network, denoted as $G_c(k) = (\mathcal{N}, \mathcal{S}, \mathcal{E}_c)$ (the top network in Fig. 5.1). In this work, a set of cooperative-state feedback gains K_{ij} will be predetermined prior to the MPC online computations, each corresponds to a G_c.

5.5.3 Perturbed Cooperative-State Feedback

The partially decentralised scheme employs the perturbed cooperative-state feedback (PSF) strategy, 'virtually', to form the stability constraints for all local MPCs of respective subsystems whose control inputs u_i, $i = 1, 2, \ldots, h$, have the following form, implicitly:

$$u_i(k) = w_i(k) + \sum_j K_{ij} x_j(k), \tag{5.18}$$

where x_j are the known neighbours' states via the available communication channels, K_{ij} is the cooperative-state feedback gain, and w_i is the perturbation variable. The PSF strategy is characterised by the following.

- The PSF is only a virtual form, and u_i will not be determined from $w_i(k)$ and $\sum_j K_{ij} x_j(k)$. Only the gain K_{ij} and neighbours' states $x_j(k)$ are assumed known. The control algorithm will not find $w_i(k)$, but $u_i(k)$ directly. The above PSF form is to merely shape the stability constraint with respect to u_i of the form $u_i(k) \in \mathbb{V}_i(k)$ for the MPC. The MPC will then compute $u_i(k)$ online incorporating the shaped stability constraint.
- The cooperative-state feedback gain K_{ij} is determined in accordance with the communication network topology, which is predefined and not necessarily identical to the fixed subsystem dynamic-coupling topology. Therefore, a set of different K_{ij} gains will be pre-computed off-line, each corresponds to a particular network topology. It is assumed in this work that all the connection topologies are known in advance.
- A feedback gain K_{ij} will be selected online following the topology changes. The neighbours' states $x_j(k)$ are also obtained online via the inter-subsystem communication links. The stability constraint $u_i(k) \in \mathbb{V}_i(k)$ for the MPC will then be derived from K_{ij} and $x_j(k)$. The MPC stability constraint is thus a cooperative constraint that is shaped by using the online information of neighbours' states $x_j(k)$ and the pre-computed K_{ij}.

- The cost function of a local MPC is, however, decoupled, i.e. only includes the local state and control variables. As a result, the computational time can be reduced substantially. The demand for data exchanged between subsystems is also reduced. The above cooperative constraint will, in addition, eliminate the conservativeness that may be caused by this local cost function, yet guarantees the global system stability. Herein, we target the applications that prefer a decentralised control architecture and readily adopt a local cost function, such as those in large-scale power systems.

For a given G_c, denote the PSF (5.18) as $u_i(k) = w_i(k) + z_i(k)$, where

$$z_i(k) = \sum_{j \in \mathcal{N}_i(G_c)} K_{ij} x_j(k). \tag{5.19}$$

The subsystem model can then be rewritten as

$$x_i(k+1) = A_i x_i(k) + B_i w_i(k) + B_i z_i(k) + E_i v_i(k), \tag{5.20}$$

where $w_i \in \mathbb{W}_i \subset \mathbb{R}^{m_i}$, and

$$v_i(k) = \sum_{j \in \mathcal{N}_i(G_s)} F_j x_j(k). \tag{5.21}$$

5.5.4 Adjacent Matrices

The subsystem dynamic-coupling network G_s is associated with the global coupling matrix H_s (adjacent matrix) describing the connection structure of G_s. The elements of H_s are either 1 or 0 only. Similarly, the communication network G_c is associated with the adjacent matrix H_c describing the connection structure (topology) of G_c. By denoting the ith block row of H_s as $H_{s[i]}$, (5.21) is rewritten as

$$v_i = H_{s[i]} F x, \tag{5.22}$$

where $x := [x_1^T \dots x_h^T]^T$ (the global state vector of Σ), $F := \text{diag}[F_j]_1^h$. And similarly, with a block-column matrix H_c and its ith block row $H_{c[i]}$,

$$z_i = [K_{ij}]_{[i]} H_{c[i]} x, \tag{5.23}$$

where $[K_{ij}]$ is the matrix having the block elements K_{ij}.

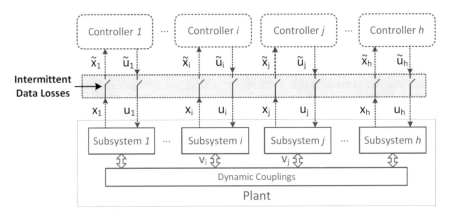

Fig. 5.3 Decentralised networked control interconnected system

5.5.5   Global System Model

For conciseness, denote $K := [K_{ij}]$. The global system Σ can then be expressed as

$$\Sigma : \ x(k+1) = A_{\Sigma} x(k) + Bw(k), \tag{5.24}$$

where $A_{\Sigma} := A + EH_s F + BKH_c$, in which

$$A := \text{diag}[A_i]_i^h, \ B := \text{diag}[B_i]_i^h, \ E := \text{diag}[E_i]_i^h, \ K := [K_{ij}],$$
$$w = [w_1{}^T \ldots w_h{}^T]^T, \ u = [u_1{}^T \ldots u_h{}^T]^T, \ w \in \mathbb{R}^m, u \in \mathbb{R}^m, x \in \mathbb{R}^n.$$

5.5.6   Deterministic Data Loss Process

In this development, a sensor/device network is installed for updating the state measurements (x_i) and actuating the control devices (u_i). It is represented by the middle block in Fig. 5.1, where the flow of signals in a decentralised control framework can be shown in Fig. 5.3. The data losses involve the following cases:

- The state vector x_i of $\mathscr{S}_i$ becomes $\tilde{x}_i$ beyond the network interface inside the respective controllers. When the state vector $x_i(k)$ at the time instant k is transmitted successfully, we have $\tilde{x}_i(k) = x_i(k)$, otherwise $\tilde{x}_i(k) = x_i(k-1)$ (i.e. when the data are lost).
- The intermittent data loss also occurs to the data exchanged between subsystems similarly to $x_i(k)$, i.e. $\tilde{x}_j(k)$ is received by its neighbours.
- For the control input (u_i), the last successfully received value of u_i will be applied to the plant during the data-lost time, i.e. $u_i(k) = u_i(k-1)$, otherwise $u_i(k) = \tilde{u}_i(k)$.

Assumption 2 The updating time instants are synchronized between all controllers of subsystems.

If we represent the consecutive updating instants of $\tilde{x}_i(k)$, $\tilde{x}_j(k)$ and $\tilde{u}_i(k)$ with a sequence of integer numbers $\mathscr{J} = \{j_1, \ldots, j_q, \ldots, j_\kappa, \ldots\}$, the time interval between j_q and j_{q+1} can be treated as one transmission period. If the communication data are perfect at time k, we have $j_\kappa = k$. The upper bound of the successful transmission period is denoted as μ (or MATI—maximum allowable transmission interval).

$$\mu := \max_{j_q \in \mathscr{J}} \left(\tau(q) \right), \quad \tau(q) := j_q - j_{q-1}. \tag{5.25}$$

The intermittent data losses are accounted for by updating the control u_i at every time step - the rolling principle of MPC. This means, the data losses are compensated for in real time, right after each incident, thanks to the MPC together with the dissipation-based constraint that also satisfies the *data-lost robust dissipative condition*. The conservativeness due to the assumption on the worst case scenario of data-lost durations, see, e.g. [82, 92] and references thereof, will thus be improved in this approach.

The *dissipation-based quadratic constraint* and the stabilisation with a *data-lost robust dissipative condition* are addressed in the next subsection.

5.5.7 Quadratic Constraint and Partially Decentralised MPC

This section introduces the dissipation-based constraint to be used with the PSF for interconnected systems that are suffered from the data-lost process characterised by (5.25).

5.5.7.1 Dissipation-Based Quadratic Constraint with PSF

Let us first define the supply rate in the perfect data environment for the controller $\mathscr{C}_i$ of node $\mathscr{S}_i$, as follows:

$$\xi_i(w_i, x_i) := w_i^T R_i w_i + 2x_i^T S_i w_i + x_i^T Q_i x_i, \tag{5.26}$$

where Q_i, R_i, S_i are time-varying coefficient matrices with symmetric Q_i, R_i. In the data-lost environment, the supply rates w.r.t. the revived time step j_q are considered. Therefore, $\xi_i(., .)$ and Q_i, R_i, S_i are denoted as $\xi_{i(q)}$ and Q_{iq}, R_{iq}, S_{iq}, at the time steps j_q, respectively. The supply rate (5.26) subject to the data-lost process μ is then rewritten as

$$\xi_{i(q)} = w_i^T R_{iq} w_i + 2x_i^T S_{iq} w_i + x_i^T Q_{iq} x_i. \tag{5.27}$$

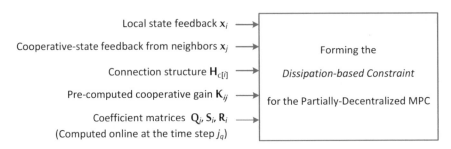

Fig. 5.4 Required information data for shaping the quadratic constraint with the PSF strategy

Definition 5.2 The input-state pair (w_i, x_i) of $\mathscr{C}_i$ is said to satisfy the *quadratic dissipativity constraint* if there exists a constant $\gamma \in \mathbb{R}^+$, $\gamma_i < 1$, such that

$$0 \geqslant \xi_{i(q)} \geqslant \gamma_i \times \xi_{i(q-1)} \ \forall j_q \in \mathscr{J}. \tag{5.28}$$

For $\xi_{i(1)} \leqslant 0$, (5.28) will be obtained by employing a convex quadratic constraint w.r.t. $u_i(j_q)$, as follows:

$$
\begin{aligned}
& w_i^T R_{iq} w_i + 2x_i^T S_{iq} w_i + x_i^T Q_{iq} x_i - \delta_{i(q-1)} \geqslant 0 \\
\Leftrightarrow \ & (u_i - z_i)^T R_{iq} (u_i - z_i) + 2x_i^T S_{iq} (u_i - z_i) - \beta_i \times \xi_{i(q-1)} \geqslant 0 \\
\Leftrightarrow \ & u_i^T R_{iq} u_i + 2\Gamma_i u_i + \psi_i \geqslant 0, \ u_i = w_i + z_i,
\end{aligned}
\tag{5.29}
$$

where $\Gamma_i := x_i^T S_{iq} - z_i^T R_{iq}$, $R_{iq} \prec 0$, $\delta_{i(q-1)} := \beta_i \times \xi_{i(q-1)}$, $0 < \beta_i < 1$,

$$\psi_i := z_i^T R_{iq} z_i - 2x_i^T S_{iq} z_i + x_i^T Q_{iq} x_i - \delta_{i(q-1)},$$

while assuming $\xi_{i(q)} \leqslant 0$.

The constraint on the current-time control vector $u_i(i_q)$ (5.29) will be imposed on the local MPC optimisation in the partially decentralised MPC scheme. The characteristics of this constraint are as follows:

- The quadratic constraint (5.29) is convex with $R_{iq} \prec 0$. For clarity, Fig. 5.4 is provided to show the information data required for shaping this quadratic constraint.
- The local state measurement x_i, the cooperative gain K_{ij} and the neighbouring states x_j are used in Γ_i and ψ_i in (5.29). The variable z_i is also calculated from Γ_i and ψ_i with the pre-computed gains K_{ij} and the online neighboring states x_j.
- The past value of the supply rate $\xi_{i(q-1)}$ is required for calculating $\delta_{i(q-1)}$ in (5.29).
- The coefficient matrices R_{iq}, S_{iq}, Q_{iq} in (5.29) are computed online at the time step j_q, or off-line, from the stabilisability condition presented in later sections. The subsystem models (A_i, B_i, E_i) and the global matrices F, H_s and H_c are required in that stabilisability condition.

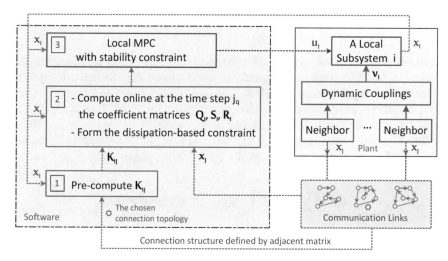

Fig. 5.5 Computational activities for forming the quadratic constraint and the MPC

5.5.8 Computation for Quadratic Constraint

With the quadratic constraint (5.29), a fully decoupled objective function without the involvement of neighbours' states can be employed in a local MPC. The computational cost is thus reduced. In exchange for that, a set of small dimension linear matrix inequalities (LMIs) will need to be solved online, at the time step j_q to update the coefficient matrices R_{iq}, S_{iq}, Q_{iq} of the quadratic constraint (5.29). Due to small dimensions, the computational cost is not a significant burden, especially for the modern computerized systems. Furthermore, the coefficient matrices R_{iq}, S_{iq}, Q_{iq} can also be pre-computed off-line in many applications. The numerical example in Sect. 5.6.2 will demonstrate these online and off-line options.

The quadratic constraint formation is shown in Fig. 5.5. In this figure, the cooperative-state feedback gains K_{ij} are firstly pre-computed off-line from the pre-determined connection topology of G_c prescribed in H_c and the node models (5.20), represented by block no. 1 on the left. The global system model will also be used in the LMI optimisation for pre-computing K_{ij}. The coefficient matrices Q_i, S_i, R_i of the quadratic constraint are then updated online at the time step j_q, if necessary (detailed in the next section). With the obtained K_{ij}, the cooperative states x_j, the updated coefficient matrices Q_{iq}, S_{iq}, R_{iq} and the past value of $\xi_i(j_{q-1})$, the quadratic constraint (5.29) is shaped, as shown in block no. 2. The MPC optimisation subsequently enforces this quadratic constraint in computing u_i online, as indicated in block no. 3. Further details are given in later sections.

5.5.9 Decentralised MPC Problem Description

The control problem here is to drive h subsystems of Σ from their initial states $x_i(0)$ (belonging to an invariant set to assure the recursive feasibility) to their zero equilibrium without any constraint violations. The control $u_i(j_q)$ will be computed online by a model predictive control algorithm that employs the stand-alone model (5.17) only, i.e. $v_i = 0$. The MPC optimisation is imposed with the quadratic constraint (5.29). It is worth emphasising here that the control u_i, but not the perturbation w_i, is involved in this optimisation. Consider the following standard objective function [88]:

$$\mathscr{J}_i(k) = \sum_{\ell=1}^{N_i} \| x_i^T(k+\ell) \|_{\mathscr{Q}_{xi}} + \| u_i^T(k+\ell-1) \|_{\mathscr{R}_{ui}},$$

where $\mathscr{Q}_{xi}, \mathscr{R}_{ui}$ are weighting (positive semi-definite) matrices, and N_i is the predictive horizon. The local optimisation problem of

$$\min_{\hat{u}_i} \mathscr{J}_i(k)$$
$$\text{subject to} \quad u_i \in \mathbb{U}_i, \quad x_i \in \mathbb{X}_i, \quad (5.17), \text{ and } (5.29) \text{ whenever } \delta_{i(k-1)} < 0, \tag{5.30}$$

is then solved by the local solver for the predictive vector sequence $\hat{u}_i := \{\hat{u}_i(k)\,\hat{u}_i(k+1) \ldots \hat{u}_i(k+N_i-1)\}$. Only the first element $\hat{u}_i^*(k)$ of the minimising sequence $\hat{u}_i^*$ which consists of N_i elements of $\hat{u}_i^*(k+\ell)$, $\ell = 0, 1, \ldots, N_i - 1$, is used to control $\mathscr{S}_i$. The whole process then repeats at the next time step.

Assumption 3 The decentralised MPC optimisations (5.30) are recursively feasible at every updating time step j_q; see, e.g. [93].

To assure the closed-loop system stability for the global system Σ (5.24) with a decentralised MPC in the presence of data losses, it may be required that the coefficient matrices Q_i, S_i, R_i are conditionally re-computed online, then the updated quadratic constraint is enforced by each local MPC optimisation. This requirement is from the stabilisability condition stated in the next section.

5.5.10 Stabilisability Condition and Control Algorithm

For the stabilisation of Σ with given graphs G_s and G_c, the input u_i is required to be bounded by the above quadratic constraint (5.29), pairing with the standard dissipation inequality [172] and the extended 'data-lost robust' dissipation inequality introduced in this section.

5.5.10.1 Convergence and Stabilisability

The stabilisability conditions for the global system Σ in the presence of intermittent data losses and inter-subsystem communication breaks will be developed in the next subsections. The coefficient matrices R_{iq}, S_{iq}, Q_{iq} of the quadratic constraint and the cooperative-state feedback gains K_{ij} will be computed from those conditions. The nominal stabilisability condition for the global system without any data losses is firstly re-stated below as a foundation for the later subsystem conditions with the intermittent data-lost process. This proposition re-states the asymptotic attractability conditions in previous chapters. The term 'stabilisability condition' is used in this chapter to address the same asymptotic attractability property. Denote $P := \text{diag}[P_i]_1^h$.

Proposition 5.1 *Consider the nominal global system Σ (5.24) without any data losses and the supply rate function $\xi(w, x) = w^T R w + 2 x^T S w + x^T Q x$. Suppose there exists a storage function $V(x) = x^T P x$, $P = P^T \succ 0$, such that for each $x_0 = x(0)$, the following conditions hold for all $k > 0$:*

$$(i) \; x^T P_m x \leqslant x^T P x \leqslant x^T P_M x, \tag{5.31}$$
$$where \; P_m \succ 0, \; P_M \succ 0, \; P_m - P_M \prec 0,$$
$$\lambda_{min}(P_m) \geqslant \phi_m > 0, \; \lambda_{max}(P_M) \leqslant \phi_M, \; \phi_m \; and \; \phi_M \; are \; finite.$$
$$(ii) \; x(k+1)^T P x(k+1) - \sigma x(k)^T P x(k) \leqslant -\xi(w(k), x(k)); \; and \tag{5.32}$$
$$(iii) \; 0 \geqslant \xi(w(k), x(k)) \geqslant \gamma \times \xi(w(k-1), x(k-1)), \tag{5.33}$$
$$for \; 0 < \gamma < 1;$$
$$(iii) \; Assumption \; 1 \; in \; Chap. \; 2 \; holds;$$

Then, the nominal system Σ (5.24) is stabilised in the sense that $x(k)$ is bounded and $x(k) \to 0$ as $k \to \infty$.

The conditions in Proposition 5.1 are for the global system. The dissipative condition for the global system Σ cannot be employed in this problem since there are also data losses in the communication network G_c. For a decentralised control scheme, it is also required to establish the stabilisability from conditions for subsystems. The quadratic constraint for a subsystem has been provided in (5.29). The dissipative condition for a controlled subsystem is now stated in the next subsection.

5.5.10.2 Data-Lost Robust Dissipative Condition

Here, a vector $\breve{g}_i$ at the time step $k + \tau$ is defined as a stacking vector of τ previous-time vectors g_i:

$$\breve{g}_i(k)^T := [g_i^T(k + \tau - 1) \; g_i^T(k + \tau - 2) \; \ldots \; g_i^T(k)]_\tau.$$

The notation $\mathsf{diag}[M_i]_1^\tau$ is a block diagonal matrix having τ matrix elements of $M_1, M_2, \ldots, M_\tau$.

For a subsystem (5.20), the *data-lost robust dissipative system* is defined w.r.t. the supply rate $\breve{\xi}_{\Delta i}$ of the following form:

$$\breve{\xi}_{\Delta i}(x_i, \breve{w}_i, \breve{z}_i, \breve{v}_i) := \breve{z}_i^T \breve{T}_i \breve{z}_i + \breve{v}_i^T \breve{X}_i \breve{v}_i + 2x_i^T (\breve{U}_i \breve{z}_i + \breve{Y}_i \breve{v}_i) + 2x_i^T Q_i x_i$$
$$+ \breve{w}_i^T \breve{R}_i \breve{w}_i + 2x_i^T \breve{S}_i \breve{w}_i + x_i^T Q_i x_i, \quad (5.34)$$

where $\breve{R}_i, \breve{T}_i, \breve{X}_i$ are symmetric, defined as

$$\breve{R}_i := \mathsf{diag}[R_i]_1^\tau, \quad \breve{T}_i := \mathsf{diag}[T_i]_1^\tau, \quad \breve{X}_i := \mathsf{diag}[X_i]_1^\tau,$$

$$\text{while} \quad \breve{S}_i := [S_i \ldots S_i]_\tau, \quad \breve{U}_i := [U_i \ldots U_i]_\tau, \quad \breve{Y}_i := [Y_i \ldots Y_i]_\tau.$$

For clarity, we split the supply rate function of $\breve{\xi}_{\Delta i}$ into two parts, as follows:

$$\breve{\xi}_{\Delta i}(x_i, \breve{w}_i, \breve{z}_i, \breve{v}_i) = \breve{\xi}_\Delta(x_i, \breve{z}_i, \breve{v}_i) + \breve{\xi}_i(x_i, \breve{w}_i), \quad (5.35)$$

where $\breve{\xi}_\Delta(x_i, \breve{z}_i, \breve{v}_i) := \breve{z}_i^T \breve{T}_i \breve{z}_i + \breve{v}_i^T \breve{X}_i \breve{v}_i + 2x_i^T (\breve{U}_i \breve{z}_i + \breve{Y}_i \breve{v}_i) + 2x_i^T Q_i x_i$, and

$\breve{\xi}_i(x_i, \breve{w}_i) := \breve{w}_i^T \breve{R}_i \breve{w}_i + 2x_i^T \breve{S}_i \breve{w}_i + x_i^T Q_i x_i.$

Also, $\xi_i(x_i, w_i) := w_i^T R_i w_i + 2x_i^T S_i w_i + x_i^T Q_i x_i.$

In what follows, we denote $\phi_{i(k)} := \breve{\xi}_{\Delta i}(x_i(k), \breve{w}_i(k), \breve{z}_i(k), \breve{v}_i(k))$ for conciseness.

Definition 5.3 Subsystem $\mathscr{S}_i$ with PSF (5.20) is said to be *data-lost robustly dissipative* with respect to the supply rate $\phi_{i(k)}$, if there exists a non-negative storage function $V_i(x_i) := x_i^T P_i x_i$, $P_i = P_i^T \succ 0$, such that for all $\breve{w}_i, \breve{z}_i, \breve{v}_i$, and all $k > 0$, the following dissipation inequality is satisfied irrespectively of the initial value of the state $x_i(0)$, for all $\tau = 1 \ldots \mu$:

$$V_i(x_i(k + \tau)) - \sigma \times V_i(x_i(k)) \leqslant \phi_{i(k)}, \ 0 < \sigma < 1. \quad (5.36)$$

Both the quadratic constraint (5.29) and the data-lost robust dissipation inequality (5.36) will be used in the stabilisability condition for interconnected systems.

Now, the data-lost robust dissipative condition with linear matrix inequalities (LMIs) for the controlled subsystem $\mathscr{S}_i$ is derived. Denote

$$\mathscr{A}_i := (A_i)^\tau,$$
$$\mathscr{B}_i := [(A_i)^{\tau-1} B_i \quad (A_i)^{\tau-2} B_i \quad \ldots \quad A_i B_i \quad B_i],$$
$$\mathscr{E}_i := [(A_i)^{\tau-1} E_i \quad (A_i)^{\tau-2} E_i \quad \ldots \quad A_i E_i \quad E_i].$$

Proposition 5.2 *Let* $0 < \sigma < 1$. *The subsystem* $\mathscr{S}_i$ *is data-lost robustly dissipative as in Definition 5.3 if the following LMI is feasible in* $P_i, Q_i, \check{S}_i, \check{R}_i,$ $\check{T}_i, \check{U}_i, \check{X}_i, \check{Y}_i, \check{Z}_i,$ *for* $P_i \succ 0$ *and* $\tau = \mu$:

$$
\begin{bmatrix}
P_i & P_i \mathscr{A}_i & P_i \mathscr{B}_i & P_i \mathscr{B}_i & P_i \mathscr{E}_i \\
* & \sigma P_i + 3Q_i & \check{S}_i & \check{U}_i & \check{Y}_i \\
* & * & \check{R}_i & 0 & 0 \\
* & * & * & \check{T}_i & 0 \\
* & * & * & 0 & \check{X}_i
\end{bmatrix} \succ 0. \tag{5.37}
$$

Proof Subsystem $\mathscr{S}_i$(5.17) is dissipative if LMI (5.37) is fulfilled since the dissipation inequality (5.36) holds true for every $x_i, \check{w}_i, \check{z}_i, \check{v}_i$. This is proved by using the basic dissipative condition in the literature, such as those in [17], plus using the prediction of $x_{i(k+\tau)}$ from $x_{i(k)}, x_i(k+\tau) = \mathscr{A}_i x_i(k) + [\mathscr{B}_i^T \ \mathscr{B}_i^T \ \mathscr{E}_i^T][\check{w}_i^T \ \check{z}_i^T \ \check{v}_i^T]^T,$ and some arrangements with the Schur compliment [14]. This dissipative condition is satisfied in the worst case of data losses when $\tau = \mu$. ∎

This dissipative condition and the quadratic constraint (5.29) will be employed in the stabilisability condition below. The stability condition is developed from the nominal stability condition in Proposition 5.1.

5.5.10.3 Stabilisability Condition

The stabilisability condition for the global system Σ with h decentralised controllers u_i having an implicit PSF form (5.18) is stated in the following theorem. The three following sub-conditions will be required:

1. Each controlled subsystem is dissipative w.r.t. $\check{\xi}_{\Delta i}(x_i, \check{w}_i, \check{z}_i, \check{v}_i)$ as defined in (5.36).
2. All h dissipation-based constraints (5.29) are satisfied.
3. An additional condition on the global interconnection structures H_s and H_c is incorporated, such that the storage function of the global system converges: $V(x) \rightarrow 0$ as $t \rightarrow +\infty$. In this work, the global storage function $V(x(k)) = \sum_{i=1}^{h} V_i(x_i(k))$, in which $V_i(x_i) = x_i^T P_i x_i$, $P_i = P_i^T \succ 0$, as in Definition 5.3, is considered.

With the bounds on the storage function $V(x)$ and the satisfaction of the three above sub-conditions, the closed-loop system stabilisability can then be established with every initial state $x(0)$.

Denote the following global matrices:

$$
\mathscr{X}_Q := F^T H_s^T \text{diag}[X_i]_1^h H_s F + F^T H_s^T \text{diag}[Y_i^T]_1^h + \text{diag}[Y_i]_1^h H_s F + \text{diag}[Q_i]_1^h,
$$

$$\mathcal{Y}_Q := \sum_{i=1}^{h} H_{c[i]}^T K_{[i]}^T T_i K_{[i]} H_{c[i]} + H_{c[i]}^T K_{[i]}^T U_i^T M_i^T + M_i U_i K_{[i]} H_{c[i]} + M_i^T Q_i M_i,$$

where $x^T M_i^T Q_i M_i x = x_i^T Q_i x_i$.

Theorem 5.3 *Consider the global system Σ (5.24) consisting of h subsystems $\mathcal{S}_i$ and the data-lost process μ (5.25). Suppose there exist h storage functions $V_i(x) = x_i^T P_i x_i$, $P_i = P_i^T \succ 0$, $i = 1, 2, \ldots, h$, such that for each $x_0 = x(0)$, the following conditions hold for all $k > 0$:*

1. *The bounds of global storage function:*

$$x^T P_m x \leqslant x^T P x \leqslant x^T P_M x, \tag{5.38}$$

   where $P = \mathsf{diag}[P_i]_1^h$, $P_m \succ 0$, $P_M \succ 0$, $P_m - P_M \prec 0$, $\lambda_{min}(P_m) \geqslant \phi_m > 0$, $\lambda_{max}(P_M) \leqslant \phi_M$, ϕ_m and ϕ_M are finite.
2. *The subsystem dissipative condition:*

$$x_i(k+\tau)^T P_i x_i(k+\tau) - \sigma_i x_i(k)^T P_i x_i(k) \leqslant \phi_{i(k)}, \text{ for } i = 1 \ldots h, \tag{5.39}$$

   where $\phi_{i(k)} := \check{\xi}_{\Delta i}\big(x_i(k), \check{w}_i(k), \check{z}_i(k), \check{v}_i(k)\big)$ defined in (5.34);
3. *The matrix inequality condition on global interconnections:*

$$\mathcal{X}_Q(X_i, Y_i, Q_i|_{i=1,2,\ldots,h}) + \mathcal{Y}_Q(T_i, U_i, Q_i|_{i=1,2,\ldots,h}) \prec 0; \tag{5.40}$$

4. *The quadratic constraint of each subsystem:*

$$0 \geqslant \xi_{i(q)} \geqslant \gamma_i \times \xi_{i(q-1)}, \ 0 < \gamma_i < 1, \text{ for } i = 1 \ldots h, \tag{5.41}$$

   where $\xi_{i(q)} := \xi_i(w_i(j_q), x_i(j_q))$ defined in (5.27);
5. *The constraint on $\xi_{i(q)}$ (5.27) and ξ_i (5.35): For $k = j_q$,*

$$\xi_{i(k)}(x_i, w_i) \leqslant \eta x_i^T Q_i x_i - \xi_{i(q)}(x_i, w_i), \tag{5.42}$$

   where $\eta = 3\frac{\mu-1}{\mu}$;
6. *Assumption 1 in Chap. 2 holds;*

   Then, the system Σ (5.24) is stabilised in the sense that $x(k)$ is bounded and $x(k) \to 0$ as $k \to \infty$, by h decoupled controls u_i, $i = 1, 2, \ldots, h$, having the PSF form (5.18).

Proof This is extended from Proposition 5.1 with the data-lost robust dissipative condition for each subsystem. Conditions (2) and (3) together ensure that the condition (2) in Proposition 5.1 is satisfied for every $x(k)$ and $w(k)$ in the presence of data losses ($V(x(k+\tau))$ replaces $V(x(k+1))$). Conditions (4) and (5) together ensure that the condition (3) in Proposition 5.1 is satisfied for all cases of data losses, and by the same token, create a more flexible condition such that the coefficient matrices Q_{iq}, S_{iq}, R_{iq}, can be adjusted online if necessary. From (5.39), (5.40), (5.41) and (5.42), we obtain the following inequalities:

$$V(x(k+\tau)) - \sigma V(x(k)) \leqslant \sum_{i=1}^{h} \xi_{\Delta i(k)}(x_i, \breve{w}_i)$$

$$= \sum_{i=1}^{h} \xi_{\Delta(k)}(x_i, \breve{z}_i, \breve{v}_i) + \sum_{i=1}^{h} \xi_{i(k)}(x_i, \breve{w}_i)$$

$$= \mu \sum_{i=1}^{h} \xi_{\Delta(k)}(x_i, z_i, v_i) + \mu \sum_{i=1}^{h} \xi_{i(k)}(x_i, w_i)$$

$$- 3(\mu - 1) \sum_{i=1}^{h} x_i^T Q_i x_i \text{ due to (5.40) and (5.42)}$$

$$\leqslant -\mu \sum_{i=1}^{h} \xi_{i(q)}(x_i, w_i) \leqslant -\mu \sum_{i=1}^{h} \gamma_i \times \xi_{i(q-1)},$$

where $V(x(k)) = x(k)^T P x(k) = \sum_{i=1}^{h} V_i(x_i(k)) = \sum_{i=1}^{h} x_i(k)^T P_i x(k)_i$, and $\sigma = \max_i(\sigma_i)$,

from which the convergence of Σ in Proposition 5.1 is obtained for the case of $\tau = 1$, and

$$\xi(x(j_q), w(j_q)) = \sum_{i=1}^{h} \xi_i(x_i(j_q), w_i(j_q)). \quad \blacksquare$$

The next theorem states the stabilisability condition in Theorem 5.3 using the LMI derived from the dissipative condition for a subsystem, and for the local control satisfaction: The quadratic constraint (5.41) can be fulfilled by $u_i(k) \in \mathbb{U}_i \subset \mathbb{R}^{m_i}$. Herein, instead of having $3Q_i$ in the LMI (5.37) we will use only Q_i, then add and subtract the term $x_i^T(Z_i - Z_i)x_i$ to the inequality, i.e.

$$\breve{\xi}_\Delta(x_i, \breve{z}_i, \breve{v}_i) = \breve{z}_i^T \breve{T}_i \breve{z}_i + \breve{v}_i^T \breve{X}_i \breve{v}_i + 2x_i^T(\breve{U}_i \breve{z}_i + \breve{Y}_i \breve{v}_i)$$

in (5.35), plus introducing a new variable Z_i, such that the conditions in Theorem 5.3 can be solved in three consecutive steps. Denote $x_{iq} := x_i(j_q)$, $z_{iq} := z_i(j_q)$,

$$\mathscr{X}_Z := F^T H_s^T \text{diag}[X_i]_1^h H_s F + F^T H_s^T \text{diag}[Y_i^T]_1^h + \text{diag}[Y_i]_1^h H_s F + \text{diag}[Z_i]_1^h,$$

$$\mathscr{X}_i(W_i, Z_i) := \mathscr{X}_Z(X_i, Y_i, Z_i|_{i=1...h}) + 2 \sum_{i=1}^{h} M_i U_i T_i^{-1} W_i H_{c[i]},$$

i.e. the decision variables in the respective LMI are W_i and Z_i, and similarly, the decision variables in $\mathcal{X}(X_i, Y_i, Z_i|_{i=1\ldots h})$ are $3 \times h$ variables of $X_i, Y_i, Z_i, i = 1, 2, \ldots, h$.

Theorem 5.4 *Let $0 < \sigma < 1$ and $\varepsilon_2 > \varepsilon_1 > 0$. Suppose that*
   (i)− $(P_i, Q_{i0}, Q_i, S_i, R_i, T_i, U_i, X_i, Y_i)$ is a feasible solution to the following off-line LMIs:

$$\mathcal{X}_Z(X_i, Y_i, Z_i|_{i=1,2,\ldots,h}) \prec 0, \tag{5.43}$$

$$\mu^{-1}Q_i + Q_{i0} - Z_i \prec 0, \ Q_{i0} \prec 0, \ and \tag{5.44}$$

$$\begin{bmatrix} P_i & P_i\mathcal{A}_i & P_i\mathcal{B}_i & P_i\mathcal{B}_i & P_i\mathcal{E}_i \\ * & \sigma P_i + Q_i & \breve{S}_i & \breve{U}_i & \breve{Y}_i \\ * & *, & \breve{R}_i & 0 & 0 \\ * & * & * & \breve{T}_i & 0 \\ * & * & * & 0 & \breve{X}_i \end{bmatrix} \succ 0, \quad P_i \succ 0, \quad \varepsilon_1 \leqslant \|P_i\| \leqslant \varepsilon_2, \tag{5.45}$$

$$i = 1, 2, \ldots h;$$

   (ii)−With those feasible Q_i, T_i, U_i, X_i, Y_i from (i), (Q_{i1}, W_i, Z_{i1}) is a feasible solution to the following off-line LMIs:

$$\begin{bmatrix} \frac{\alpha_i}{h}\mathcal{X}_i(W_i, Z_{i1}) & H_{c[i]}^T W_i^T \\ * & -T_i \end{bmatrix} \prec 0, \ and \tag{5.46}$$

$$\mu^{-1}Q_i + Q_{i1} - Z_{i1} \prec 0, \ Q_{i1} \prec 0, \tag{5.47}$$
$$i = 1, 2, \ldots h,$$

in which $\sum_{i=1}^{h} \alpha_i = h;$
   (iii)−With those feasible Q_i, S_i, R_i from (i) and Z_{i1} from (ii), (R_{iq}, S_{iq}, Q_{iq}), is a feasible solution to the following online LMIs whenever $\psi_i(j_q) < 0$:

$$\begin{bmatrix} R_i + R_{iq} & (S_i^T + S_{iq}^T)x_{iq} \\ * & x_{iq}^T(\mu^{-1}Q_i + Q_{iq} - Z_{i1})x_{iq} \end{bmatrix} \preccurlyeq 0, \ Q_{iq} \prec 0, \tag{5.48}$$

$$z_{iq}^T R_{iq} z_{iq} + 2x_{iq}^T S_{iq} z_{iq} + x_{iq}^T Q_{iq} x_{iq} - \delta_{i(q-1)} \geqslant 0, \ R_{iq} \prec 0, \tag{5.49}$$
$$i = 1, 2, \ldots, h;$$

   (iv)−Assumption 3 holds true;
   Then, any h optimising controls from (5.30), $\hat{u}_i^*(j_q) \in \mathbb{U}_i, 0 \in \mathbb{U}_i$, for $i = 1, 2, \ldots, h$, having the virtual PSF form (5.18) with $[K_{ij}]_{[i]} \equiv T_i^{-1}W_i$, that fulfil the

quadratic constraints (5.29), stabilize Σ (5.24) and (5.25) in the sense that $x(k)$ is bounded, and $x(k) \to 0$ as $k \to \infty$.

Proof (1) *Data-Lost Robust Dissipativity*: From Proposition 1, all subsystems $\mathscr{S}_i$, $i = 1 \ldots h$, are data-lost robust dissipative due to (5.45) with

$$\check{\xi}_\Delta(x_i, \check{z}_i, \check{v}_i) = \check{z}_i^T \check{T}_i \check{z}_i + \check{v}_i^T \check{X}_i \check{v}_i + 2x_i^T (\check{U}_i \check{z}_i + \check{Y}_i \check{v}_i) \qquad (5.50)$$

in (5.36).

(2) *Asymptotic attractivity*: Conditions (1) and (2) in Theorem 5.3 are satisfied by (5.45) with $\check{\xi}_\Delta(x_i, \check{z}_i, \check{v}_i)$ in (5.50). Condition (3) in Theorem 5.3 is satisfied by (5.43) and (5.46). Condition (5) in Theorem 5.3 is satisfied by (5.48). The LMIs (5.44) and (5.47) will ensure that (5.48) is feasible. The quadratic constraint (5.29) will have at least a solution of $u_i = 0$ due to (5.49), which ensures that Condition (4) in Theorem 5.3 is fulfilled (the condition (5.49) in (iii) ensures that the ellipsoid in u_i defined by the inequality (5.29) does not reduce to a singleton, and $u_i = 0$ belongs to that ellipsoid, i.e. $u_i = 0$ satisfies (5.29)). Furthermore, by denoting $V(x) := \sum_{i=1}^{h} V_i(x_i)$ and

$$\xi_{(q)} := \sum_{i=1}^{h} \xi_{i(q)},$$

it follows from (5.29) and (5.36) with $\check{\xi}_\Delta(x_i, \check{z}_i, \check{v}_i)$ in (5.50), which is fulfilled by (5.45), that

$$V(x(j_{q+1})) - \sigma \times V(x(j_q)) \leqslant \vartheta_{j_q} + \beta \times |\xi_{(q-1)}|,$$

where $\beta := \mu \max_{i=1 \ldots h} \beta_i$ and

$$\vartheta_{j_q} := \mu \sum_{i=1}^{h} \begin{bmatrix} w_i(j_q) \\ x_i(j_q) \end{bmatrix}^T \begin{bmatrix} R_i + R_{iq} & S_i + S_{iq} \\ * & \mu^{-1} Q_i + Q_{iq} - Z_i \end{bmatrix} \begin{bmatrix} w_i(j_q) \\ x_i(j_q) \end{bmatrix}$$
$$+ \mu x^T(j_q) \big[F^T H_s^T \mathrm{diag}[X_i]_1^h H_s F + 2\mathrm{diag}[Y_i]_1^h H_s F + \mathrm{diag}[Z_i]_1^h$$
$$+ 2 \sum_{i=1}^{h} M_i U_i T_i^{-1} W_i II_{c[i]} + \sum_{i=1}^{h} H_{c[i]}^T W_i^T T_i^{-1} W_i H_{c[i]} \big] x(j_q),$$

similarly to the proof of Theorem 5.3 with Assumption 2.

Then, from (5.48), (5.45), (5.43), (5.46) and (5.47) (by applying the Schur complement), and with $[K_{ij}]_{[i]} \equiv T_i^{-1} W_i$, the variable ϑ_{j_q} is only negative or non-positive, the above inequality thus reads

$$V(x(j_{q+1})) \leqslant \sigma \times V(x(j_q)) + \beta \times |\xi_{(q-1)}|.$$

Similarly to the proof in [156], it is concluded that the storage function $V(.) \to 0$ as $\xi(.) \to 0$. With the conditions on the feasibility of the quadratic constraint together with the MPC recursive feasibility in Assumption 3, the convergence will be achieved for each $x(0)$. The proof is complete. ∎

The conditions in Theorem 5.4 are feasible LMI conditions with several solutions. The following corollary provides the conditions with LMI optimisations for assuring a solution. The conditions in Theorem 5.4 will not be changed, but the LMIs will become constraints of the optimisation sub-problems newly introduced. The objective function $x_i^T Q_i x_i$ has been chosen in these LMI optimisations to obtain the quadratic constraint (5.29) more likely to be feasible having at least a solution of $0 \in \mathbb{U}_i$.

Corollary 5.1 *Let $0 < \sigma < 1$. Suppose that*
(i)— *P_i, Q_i, S_i, R_i, T_i, U_i, X_i, Y_i solve the following off-line LMI optimisation:*

$$\max_{P_i, Q_{i0}, Q_i, S_i, R_i, T_i, U_i, X_i, Y_i} \quad \min_{i=1...h} \quad x_i^T(0) Q_{i0} x_i(0) \tag{5.51}$$

subject to $x_i^T(0)[\mu^{-1} Q_i + Q_{i0} - Z_i] x_i(0) \leqslant 0$, $Q_{i0} \prec 0$, (5.43), and (5.45).
(ii)—*With the optimising Q_i, T_i, U_i, X_i, Y_i from (i), Q_{i1}, W_i, Z_{i1} solve the following off-line LMI optimisation:*

$$\max_{Q_{i1}, W_i, Z_{i1}} \quad \min_{i=1...h} \quad x_i^T(0) Q_{i1} x_i(0) \tag{5.52}$$

subject to $x_i^T(0)[\mu^{-1} Q_i + Q_{i1} - Z_{i1}] x_i(0) \leqslant 0$, $Q_{i1} \prec 0$, and (5.46).
(iii)—*With the optimising Q_i, S_i, R_i from (i) and Z_{i1} from (ii), R_{iq}, S_{iq}, Q_{iq} solve the following online LMI optimisation whenever $\psi_i(j_q) < 0$:*

$$\max_{R_{iq}, S_{iq}, Q_{iq}} \quad x_{iq}^T Q_{iq} x_{iq} \tag{5.53}$$

subject to (5.48) and (5.49);
(iv)—*Assumption 3 holds true;*
Then, any h optimising controls from (5.30), $\hat{u}_i^*(j_q) \in \mathbb{U}_i$, $0 \in \mathbb{U}_i$, for $i = 1, 2, \ldots, h$, having the virtual PSF form (5.18) with $[K_{ij}]_{[i]} \equiv T_i^{-1} W_i$, that fulfil the quadratic constraints (5.29), stabilize Σ (5.24) and (5.25) in the sense that $x(k)$ is bounded, and $x(k) \to 0$ as $k \to \infty$.

Proof This is a direct result of Theorem 5.4 with the LMI optimisations (5.51), (5.52) and (5.56) are employed for the LMIs in Theorem 5.4. The objectives $x_i^T(0) Q_{i0} x_i(0)$, $x_i^T(0) Q_{i1} x_i(0)$ and $x_{iq}^T Q_{iq} x_{iq}$ are chosen for (5.51), (5.52) and (5.56), respectively, for the purpose of creating a favourable precondition for $\psi_i(j_q) > 0$ in (iii) to hold such that the quadratic constraint will be fulfilled. ∎

5.5.10.4 Enhanced Constraint Feasibility

According to the decentralised MPC optimisation problem (5.30), the control u_i must meet two conditions of $u_i \in \mathbb{U}_i$ and the quadratic constraint (5.29), which can be translated to the condition of having u_i belong to the intersection of the two sets: $\mathbb{U}_i$ and an ellipsoidal region governed by the quadratic constraint (5.29). This ellipsoid will have both the volume and the center point vary following the update of local and neighbouring states and the past supply rate. Theorem 5.4 has only considered the case of $\psi_i < 0$ in the conditions in (iii). Here, we extend the condition to cover the case of $\psi_i \geq 0$, as in the next theorem.

Theorem 5.5 *Let $0 < \sigma < 1$ and $0 < \varepsilon_i \ll \|x_i(0)\|_2$. Suppose that P_i, Q_i, S_i, R_i, T_i, U_i, X_i, Y_i are obtained off-line from items (i) and (ii) in Corollary 5.1, the variables $\mathcal{R}_{iq} := \{R_{iq}, S_{iq}, Q_{iq}\}$, λ_i and ζ_i solve the following online LMI optimisation:*

(a)−*If $\psi_i(j_q) < 0$ and $\|x_i(j_q)\|_2 > \varepsilon_i$:*

$$\max_{\mathcal{R}_{iq}, \; \lambda_i > 0, \; \zeta_i \geq \psi_i(j_q)} \quad \zeta_i \tag{5.54}$$

$$\text{subject to} \quad \begin{bmatrix} R_{iq} + \lambda_i I_{m_i} & S_{iq} x_{iq} \\ * & -\lambda_i \times \eta_i + \zeta_i \end{bmatrix} \prec 0, \; \text{and (5.48)}, \tag{5.55}$$

$$i = 1 \ldots h;$$

(b)−*If $\psi_i(j_q) < 0$ and $\|x_i(j_q)\|_2 \leq \varepsilon_i$:*

$$\max_{\mathcal{R}_{iq}} \quad x_{iq}^T Q_{iq} x_{iq} \tag{5.56}$$

subject to (5.48) and (5.49);

(c)−*Assumption 3 holds true;*

Then, any h optimising controls from (5.30), $\hat{u}_i^*(j_q) \in \mathbb{U}_i$, $0 \in \mathbb{U}_i$, for $i = 1, 2, \ldots, h$, having the virtual PSF form (5.18) with $[K_{ij}]_{[i]} \equiv T_i^{-1} W_i$, that fulfil the quadratic constraints (5.29), stabilize Σ (5.24) and (5.25) in the sense that $x(k)$ is bounded, and $x(k) \to 0$ as $k \to \infty$.

Proof There is a constrained $u_{iq} \in \mathbb{U}_i$ feasible to the quadratic constraint (5.29) if the condition in (b) is fulfilled, similarly to the condition (iii) in Corollary 5.1, since the solution to the optimisation (5.54) will make $\psi_i \geq 0$. The region of u_i that satisfies both $u_i \in \mathbb{U}_i$ and the quadratic constraint (5.29) is then further enlarged from those obtained in (b) with the condition in (a) to allow for $\psi_i < 0$. The optimisation (5.54) in (a) aims to maximize the intersection of $\mathbb{U}_i$ and the ellipsoid region governed by the quadratic constraint (5.29) while accepting $\psi_i < 0$. ■

Remarks:

1. A tuning number $\varsigma \geqslant 0$ can be used in (5.43):

$$\mathscr{X}(X_i, Y_i, Z_i|_{i=1...h}) \prec -\varsigma I,$$

   to improve the numerical tractability.
2. $S_{iq} \equiv 0$ in (5.56) may be chosen when $\|x_i\|_2$ is small ($\|x_{iq}\|_2 \leqslant \varepsilon_i$).
3. The coefficient matrix P_i of the storage function in (5.37) can also be pre-computed using the discrete-time Riccati equation [88] and the weighting matrices $\mathscr{Q}_{xi}, \mathscr{R}_{ui}$ in the MPC objective function (5.30). This means P is known in (5.37), but not a decision variable. This option has been selected in the numerical example in the next section.
4. The dissipation inequality used in the presenting approach does not have the stage cost of the MPC optimisation as the supply rate, thus, will allow for user-chosen stage-cost weighting matrices, which is preferable in practice. The user-chosen weighting matrices are influenced by other operational performance indices.
5. The condition for applying the quadratic constraint to the MPC optimisation (5.30) will need to be modified with $\psi_i(j_q) \geqslant 0$ and $\delta_{i(q-1)} < 0$ if the optimiser for the quadratically constrained quadratic program (QCQP) does not accept $\psi_i(j_q) < 0$. This only depends on the software toolbox for solving the QCQP problem.

Using the results from Theorem 5.5, a control procedure for the decentralised model predictive control is proposed in the next subsection.

5.5.10.5   Control Procedure

On the grounds of the stabilisability conditions in the above subsections, the control procedure is as follows:

1. *The PSF gains $[K_{ij}]_{[i]}$ of all subsystem $\mathscr{S}_i, i = 1, 2, \ldots, h$, are determined off-line with the predefined $H_c(s)$ from feasible solutions to LMI optimisations in (i) and (ii) of Corollary 5.1.*
2. *The time-dependent coefficients R_{iq}, S_{iq}, Q_{iq} of the quadratic constraint can be determined from the feasible solution to the LMI optimisation in either (a) or (b) of Theorem 5.5 at every updating time step j_q, if necessary. Otherwise, the off-line pre-computed matrices in Step 1, R_i, S_i, Q_i, can be employed in the quadratic constraint (5.29).*
3. *With the obtained R_{iq}, S_{iq}, Q_{iq} from the above Step 2, or R_i, S_i, Q_i from Step 1, the quadratic stability constraint (5.29) is formed using the PSF gains $[K_{ij}]_{[i]}$, the known values of x_{iq}, z_{iq} and $\delta_{i(q-1)}$ for the decentralised MPC optimisation problem (5.30).*
4. *The local MPC optimisation (5.30) is then solved for the minimizing vector sequence $\hat{\mathbf{u}}_i^* := \{\hat{u}_i^*(j_q)\ \hat{u}_i^*(j_q + 1) \ldots \hat{u}_i^*(j_q + N_i - 1)\}$, of which only the first element $\hat{u}_i^*(j_q)$ is output to control $\mathscr{S}_i$ at the revived time step j_q, as described in subsection 5.5.9.*
5. *Return to Step 2.*

5.6 Numerical Simulation

Two numerical examples are given in this section for illustration.

5.6.1 Network Process for Bauxite Ore Treatment

An interconnected system having recirculations for bauxite ore treatment in an aluminium extraction is deployed in this numerical simulation. This is the model of a preheating and pre-desilication process that has moderate interactions between dynamically coupled subsystems. The process description is as follows: the extracted bauxite ores are fed into grinding machines, then preheated and circulated through a system of settling and pre-desilicating tanks before being mixed with heated soda in the digesters. The process is divided into three subsystems. The process diagram is given in Fig. 5.6. The mill grinding area (subsystem 1) has one grinding mill (ML-1) built with a belt weigher (W-11), a mill slurry tank (T-11) and a relay tank (T-21). The heating and pre-desilication area (subsystem 2) has one steam heater (H-21) built with mill slurry pumps (P-21), booster pumps (P-22) and three pre-desilicating tanks (T-21, -22, -23) built with transfer and recirculation pumps (P-23, -24, -25). The saturated liquor from the reactor tank R-31 in subsystem 3 is extracted and recirculated back to the mill grinding area. The control input of subsystem 1 is the mill feed solid flow rate via the belt weigher (W-11). The levels in the mill, slurry and relay tanks are its states. The control inputs of pre-desilication tanks in subsystem 2 are the transfer and recirculation flow rates (P-23, -24, -25). The levels in three pre-desilication tanks are its states. The two control inputs of subsystem 3 are steam and caustic soda flow rates feeding to the top of the reactor tank R-31. The level and temperature in the reactor R-31 are its states.

The following four simulations are set up to demonstrate the PSF effectiveness:

(1) Fully decentralised MPC without any attractivity constraints;

(2) Fully decentralised MPC with dQDC-based attractivity constraint;

(3) Partially decentralised MPC with dQDC-based attractivity constraint and PSF for $H_c = H_s$;

(4) Partially decentralised MPC with dQDC-based attractivity constraint and PSF for $H_c \neq H_s$, $\mathcal{E}_c \subset \mathcal{E}_s$, specifically, $H_{c[2]} = [0\ 0\ 0]$. Each will be simulated with three different sampling times, $T_s = 0.15, 0.25$ and 0.45. In this example, the closed-loop system Σ is not stable with a centralised MPC even when the dQDC-based attractivity constraint is applied for the case of $T_s = 0.55$. MATI $= 7$ in this example. The same pattern of data losses has been used with different sampling times for different simulations. The system configuration matrices are given below.

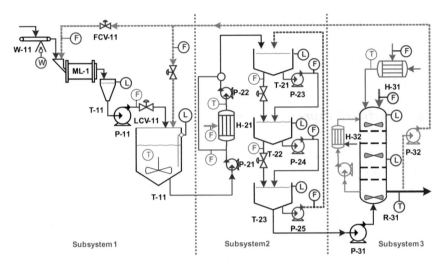

Fig. 5.6 Pre-desilication process

$$A_1 = \begin{bmatrix} -1.4 & 0.3 & 0 \\ 0 & -1.8 & 1.5 \\ 0.1 & -2.7 & 1.06 \end{bmatrix}, A_2 = \begin{bmatrix} -.76 & 0 & 0.25 \\ 0.48 & -0.56 & 0 \\ 0 & 0.2 & -0.34 \end{bmatrix},$$

$$A_3 = \begin{bmatrix} -4.3 & 5.9 \\ -1.8 & 2.7 \end{bmatrix}, B_1 = \begin{bmatrix} 0 & 0 & 1 \end{bmatrix}^T, B_2 = -I_3, B_3 = I_2,$$

$$L1 = 0.05 \begin{bmatrix} 1 & 1 & 1 \end{bmatrix}^T, L2 = 0.02 \, I_3, L3 = 0.03 \, I_2,$$

$$E_1 = \begin{bmatrix} 2.03 & 0 & 1.98 \end{bmatrix}^T, E_2 = \begin{bmatrix} 2.06 & 0 & 0 \end{bmatrix}^T, E_3 = \begin{bmatrix} 3.97 & 0 \end{bmatrix}^T$$

$$F_1 = \begin{bmatrix} 1 & 0 & 0 \end{bmatrix}, F_2 = \begin{bmatrix} 0 & 0 & 1 \end{bmatrix}, F_3 = \begin{bmatrix} 1 & 0 \end{bmatrix}, C_i = I, H_s = \begin{bmatrix} 0 & 0 & 1 \\ 1 & 0 & 0 \\ 0 & 1 & 0 \end{bmatrix}.$$

The predictive horizons $N_i = 6$, $i \in \{1, 2, 3\}$, have been chosen for all local MPCs. The control constraints $\|u_i\| \leqslant 2$. The thresholds of $\|x_i\|$ are chosen as $\varepsilon_1 = 0.02$, $\varepsilon_2 = 0.02$, $\varepsilon_3 = 0.05$, for use in the condition (b) of Theorem 5.5. The initial states of $x_1(0) = [4.35 \ -5.02 \ -6.08]^T$, $x_2(0) = [-4.06 \ 2.13 \ -1.02]^T$, $x_3(0) = [1.02 \ -1.141]^T$, and $\gamma_i = 0.9999$, $i = 1, 2, 3$ have been set herein.

The simulation results for the intermittent data losses of MATI = 6 are given in Figs. 5.7, 5.8, 5.9 and 5.10, which are interpreted as follows: The dQDC-based attractivity constraints have helped come up with the better control performances in Figs. 5.8, 5.9 and 5.10, compared to those without the dQDC-based attractivity constraints in Fig. 5.7. Among these figures, those in Fig. 5.8 for the case of PSF and $H_c = H_s$ have shown better control performances (i.e. having the least control deviations) in the lower sampling rate ranges. There are two positive results to be highlighted here. Firstly, the fully decentralised MPC delivered the best control per-

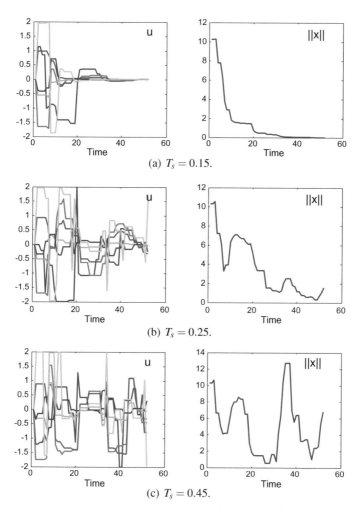

Fig. 5.7 Decentralised MPC without dQDC for NCs with MATI $= 6$

formance for the case of $T_s = 0.45$, as shown in Fig. 5.10c. This can be explained by the effect of isolations between interconnection subsystems that helps reduce the interactions of data losses among them. Secondly, the best control performance for the case of $T_s = 0.25$ is obtained when $H_c \neq H_s$, as in Fig. 5.9b. This demonstrates that the partial decentralised MPC has, indeed, delivered a supreme control performance within the moderate sampling time ranges. The remaining control performances in descending order are as follows: The partially decentralised MPC with dQDC-based constraint in Figs. 5.8 and 5.9, the fully decentralised MPC with dQDC-based constraint in Fig. 5.10, then the fully decentralised MPC without dQDC-based constraint in Fig. 5.7.

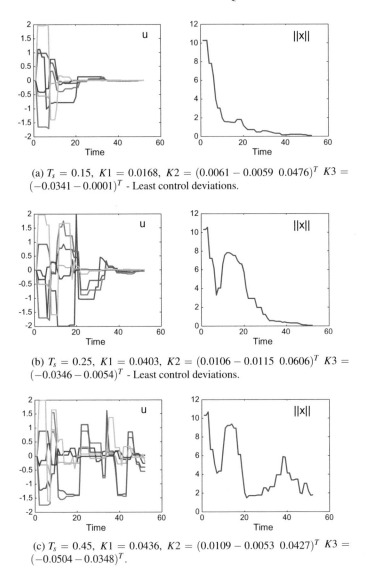

(a) $T_s = 0.15$, $K1 = 0.0168$, $K2 = (0.0061 - 0.0059\ 0.0476)^T$ $K3 = (-0.0341 - 0.0001)^T$ - Least control deviations.

(b) $T_s = 0.25$, $K1 = 0.0403$, $K2 = (0.0106 - 0.0115\ 0.0606)^T$ $K3 = (-0.0346 - 0.0054)^T$ - Least control deviations.

(c) $T_s = 0.45$, $K1 = 0.0436$, $K2 = (0.0109 - 0.0053\ 0.0427)^T$ $K3 = (-0.0504 - 0.0348)^T$.

Fig. 5.8 Decentralised MPC with dQDC and PSF, $H_c = H_s$, for NCs with MATI $= 6$

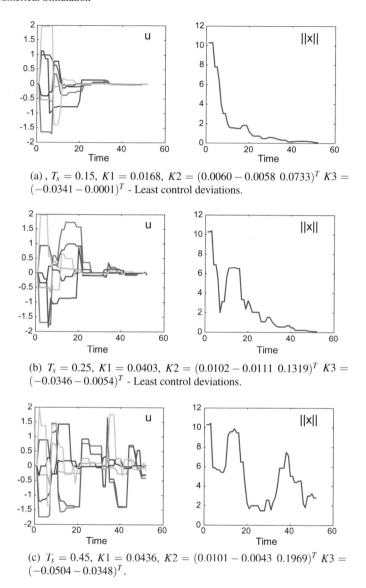

(a) , $T_s = 0.15$, $K1 = 0.0168$, $K2 = (0.0060 - 0.0058 \; 0.0733)^T$ $K3 = (-0.0341 - 0.0001)^T$ - Least control deviations.

(b) $T_s = 0.25$, $K1 = 0.0403$, $K2 = (0.0102 - 0.0111 \; 0.1319)^T$ $K3 = (-0.0346 - 0.0054)^T$ - Least control deviations.

(c) $T_s = 0.45$, $K1 = 0.0436$, $K2 = (0.0101 - 0.0043 \; 0.1969)^T$ $K3 = (-0.0504 - 0.0348)^T$.

Fig. 5.9 Decentralised MPC with dQDC and $H_c \neq H_s$, $(H_{c[2]} = [0 \; 0 \; 0])$ for NCS with MATI $= 6$

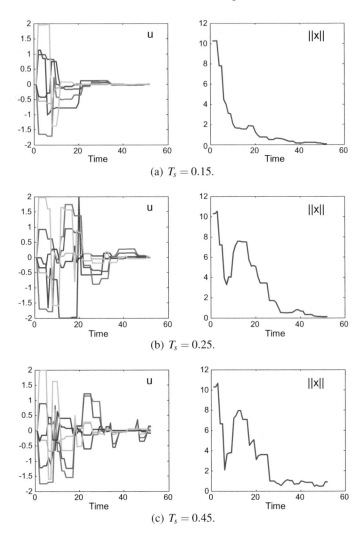

(a) $T_s = 0.15$.

(b) $T_s = 0.25$.

(c) $T_s = 0.45$.

Fig. 5.10 Decentralised MPC with dQDC and $H_c = 0$ for NCS with MATI $= 6$

Intensive simulation for each case of attractivity constraints has also been studied with over 20 different data-lost patterns. The general observation is as follows:

(i) The control moves with the dQDC-based attractivity constraint in the case of no data losses are, in average, slightly more conservative than the MPC alone without the attractivity constraint.

(ii) In exchange for that, the deviations of control moves are exceeding when the sampling time T_s increases and if the dQDC-based attractivity constraint is not employed in the local MPC optimisations.

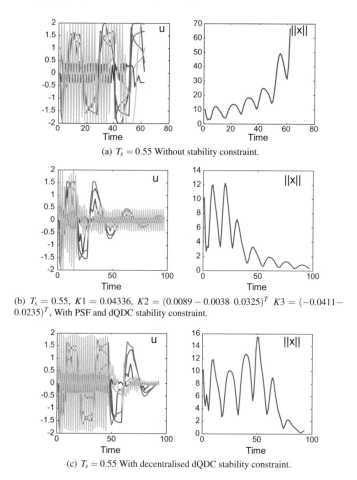

(a) $T_s = 0.55$ Without stability constraint.

(b) $T_s = 0.55$, $K1 = 0.04336$, $K2 = (0.0089 - 0.0038\ 0.0325)^T$ $K3 = (-0.0411 - 0.0235)^T$, With PSF and dQDC stability constraint.

(c) $T_s = 0.55$ With decentralised dQDC stability constraint.

Fig. 5.11 Decentralised MPC with dQDC in a perfect data environment

(**iii**) However, the state deviations with the dQDC-based attractivity constraint are slightly higher than those without the constraint at the lower ends of T_s.

(**iv**) Also, the control moves in the case of dQDC with PSF are almost identical to those without dQDC at the very low ends of T_s when there are no data losses.

(**v**) Nevertheless, there are some cases of the data-lost pattern (around 4 out of 20 simulation runs), the above observations are not true. The control moves may actually have less deviations than those with the dQDC-based attractivity constraints in these minority cases.

For demonstrating the stabilisability aspect of the dQDC-based attractivity constraint, simulation results for $T_s = 0.55$ without data losses have been obtained and reported in Fig. 5.11. The system is clearly unstable without the attractivity constraint. Simulation results for $T_s = 0.12$ are provided in Fig. 5.12, indicating the

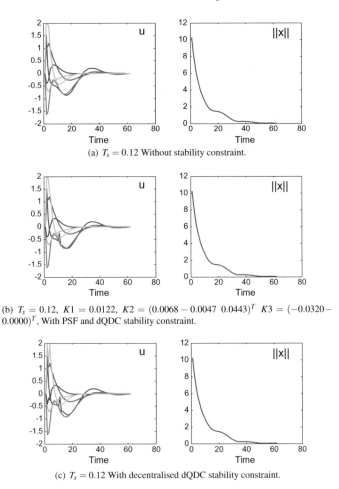

(a) $T_s = 0.12$ Without stability constraint.

(b) $T_s = 0.12$, $K1 = 0.0122$, $K2 = (0.0068 - 0.0047\ 0.0443)^T$ $K3 = (-0.0320 - 0.0000)^T$, With PSF and dQDC stability constraint.

(c) $T_s = 0.12$ With decentralised dQDC stability constraint.

Fig. 5.12 Decentralised MPC with dQDC for NCS

stabilised trajectories. Herein, the control moves in the cases of dQDC with PSF are almost identical to those without the dQDC.

The simulation studies in this section have illustrated the virtual PSF strategy with the dQDC-based attractivity constraint for the decentralised model predictive networked control of interconnected systems presented in this chapter. The second example studies the virtual PSF strategy for the automatic generation control problem with the partially decentralised model predictive networked control algorithm.

5.6.2 Automatic Generation Control of a Power System

The AGC problem in Chap. 4 is considered in this realistic example. The numerical simulation has been conducted with the following model parameters and settings: Sampling and updating time: $T_s = 2$. Initial state vectors: $x_1 = 0$; $x_2 = 0$; $x_3 = 0$; $x_4 = 0$. State constraint: $(-\infty, +\infty)$. Control constraint: $|u_i(k)| \leqslant 0.5$. Weighting coefficients: $W_x = \text{diag}\{50\,50\,0\,0\,50\,50\,0\,0\,50\,50\,0\,0\,50\,50\,0\,0\}$, $W_u = \text{diag}\{1\,1\,1\,1\}$. Predictive horizon: $N = 4$. In this example, the load in area 2 increases 25% while the load in area 3 decreases 25% from the time step 2 onward. The weighting coefficients for the two state variables ΔP_{mech} and ΔZ are set to 0, similarly to those in [167]. The set points of the MPC are assumed known in a priori here. The coefficient matrices of the quadratic constraint are either pre-computed off-line (i.e. $Q_{iq} = -Q_i$, $S_{iq} = -S_i$, $R_{iq} = -R_i$) or updated online and employed in the DeMPC in the simulations. The trajectories of state and control elements from the centralised MPC and DeMPC are shown in the respective figures below.

The centralised MPC with the predictive horizon $N = 12$ provides a desirable performance and is the benchmark for the DeMPC. However, the centralised MPC with $N = 4$ is unstable without any additional constraints as shown in Fig. 5.13b. The DeMPC with all local MPCs having the same predictive horizon $N = 4$ is not unstable, but provides a much worse control performance compared to the centralised MPC with $N = 12$. When the DeMPC is implemented with the quadratic quadratic constraint developed in this chapter, but without the PSF strategy, the time responses in Fig. 5.14 show a control performance comparable to the centralised MPC having a longer predictive horizon $N = 12$, as in Fig. 5.13a. The settling time with DeMPC is only 15% longer approximately. It is emphasised here that there are not any exchanged data between the control areas in a DeMPC scheme. In this fully decentralised case, the tie-line power steady states are quite high, as shown in Fig. 5.14, similarly to the centralised MPC with a longer predictive horizon, as shown in Fig. 5.13a.

5.6.2.1 PSF Strategy in a Perfect Data Environment

We now implement the cooperative-state feedback and the PSF strategy developed in this chapter to this AGC problem.

The control performances have been improved with the smaller tie-line power steady states in both cases, one with the full cooperative-state feedbacks from all neighbouring control areas in Fig. 5.15a, and one with only a part of the cooperative-state feedbacks in Fig. 5.15b. The first case study shown in Fig. 5.15a has the cooperative states from the control area 2 to the control area 1, from the control areas 1 and 3 to the control area 2, from the control areas 2 and 4 to the control area 3 and from the control area 3 to the control area 4. The second case study shown in Fig. 5.15b has the cooperative states from the control area 2 to the control area 1, from the control areas 1 and 3 to the control area 2 and from the control area 3 to the control area

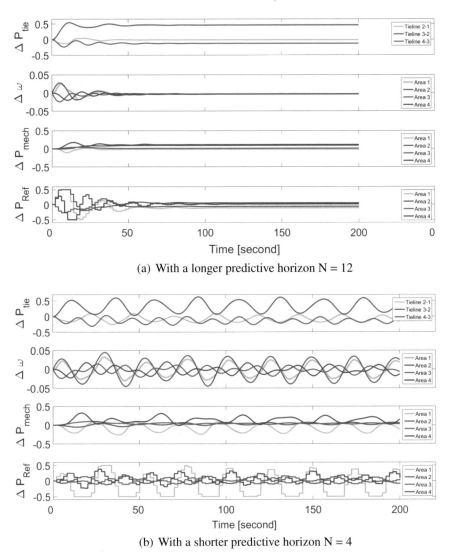

(a) With a longer predictive horizon N = 12

(b) With a shorter predictive horizon N = 4

Fig. 5.13 Centralised MPC without the quadratic constraint

4 only. The tie-line power steady states and the settling time in the second case are slightly worse than those in the first case, but is better than those in the fully DeMPC without any cooperative states, as shown in Fig. 5.14. The worst case with the PSF strategy is obviously the fully decentralised MPC in Fig. 5.14 above.

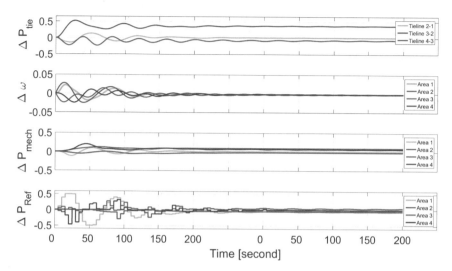

Fig. 5.14 Decentralised MPC with quadratic constraint, shorter predictive horizon N = 4. The control performance is comparable to that of the centralised MPC with $N = 12$ in Fig. 5.13a

5.6.2.2   Simulation Results with the PSF Strategy in an Intermittent Data-Lost Environment

The settling time will become longer in all cases when there are measurement data losses.

The data-lost process has been simulated with random-walk variables for the second case study with the full cooperative-state feedbacks shown in Fig. 5.15a. The resulting time responses when the PSF strategy is implemented for the DeMPC are shown in Figs. 5.16 and 5.17 for two different data-lost scenarios. The corresponding trajectories of the global state vectors that are obtained by the DeMPC after the data are lost are shown in Figs. 5.16b and 5.17b for the two cases, respectively. The system is clearly stabilised in the presence of the data-lost process of maximum 6 time steps, but with a longer settling time compared to that in Fig. 5.15. The system is unstable with the centralised MPC without any additional constraints in the presence of data losses, as expected.

5.6.2.3   Dissipation-Based Constraint with Online-Updated Coefficient Matrices

In the last figure, Fig. 5.18, the quadratic constraint has the coefficient matrices Q_{iq}, S_{iq}, R_{iq} updated online as in Step 2 of the control procedure above. The resulting time responses pose a similar performance to those in the simulations using the pre-computed coefficient matrices. It is, therefore, a problem-dependent

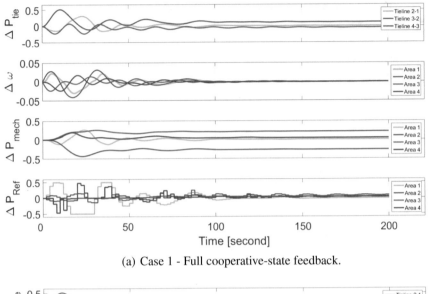

(a) Case 1 - Full cooperative-state feedback.

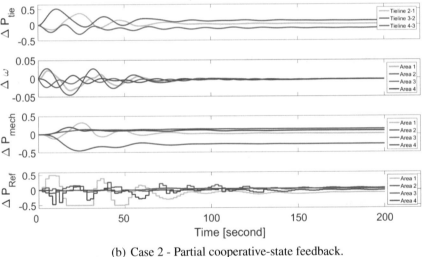

(b) Case 2 - Partial cooperative-state feedback.

Fig. 5.15 Partially decentralised MPC with PSF and quadratic constraint, shorter predictive horizon $N = 4$—two case studies

decision to apply the pre-computed coefficient matrices for the quadratic constraint or to update them online at every reviving time step.

In summary, simulation results with a power system example in this section have demonstrated the success of the virtual PSF strategy and quadratic constraints for the partially DeMPC of interconnected systems. The control performances of the partially DeMPC with quadratic constraints in various cooperative-state feedback

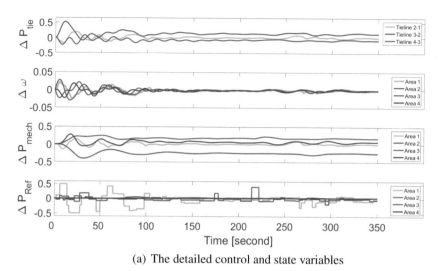

(a) The detailed control and state variables

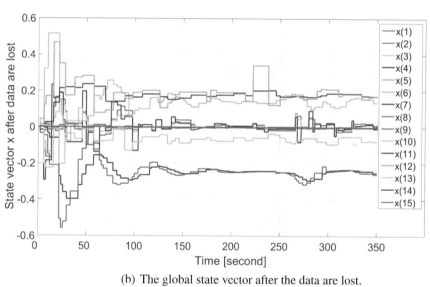

(b) The global state vector after the data are lost.

Fig. 5.16 Case 3A—Full cooperative-state feedback and intermittent data losses, $\mu = 7$. Partially decentralised MPC with PSF and quadratic constraint, $N = 4$

scenarios are comparable with those from the distributed MPC (DMPC) scheme in [167]. The DeMPC with PSF strategy can guarantee the global system stability when all the communication links that are required permanently in the DMPC scheme in [167] are lost, as well as in the presence of measurement data losses. Furthermore, the control performance of the DMPC in [167] will be established by the DeMPC with PSF strategy whenever those communication links are established.

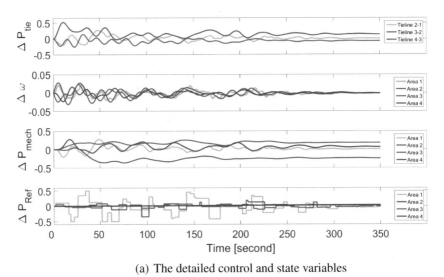

(a) The detailed control and state variables

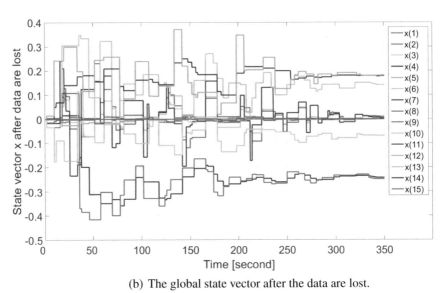

(b) The global state vector after the data are lost.

Fig. 5.17 Case 3B—Full cooperative-state feedback and intermittent data losses, $\mu = 7$. Partially decentralised MPC with PSF and quadratic constraint, $N = 4$

If the communication links are lost, the system is still assuredly stable even when there are measurement data losses (while not obtained in [167]) and, at the same time, achieves a satisfactory control performance comparable to a well-performed centralised MPC.

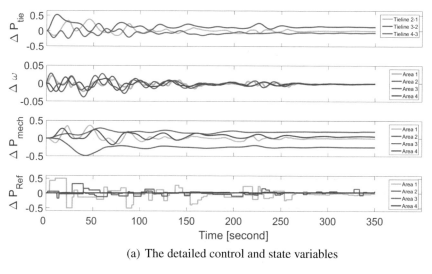

(a) The detailed control and state variables

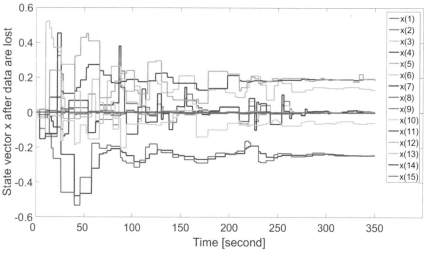

(b) The global state vector after the data are lost.

Fig. 5.18 Case 3C—Full cooperative-state feedback and intermittent data losses, $\mu = 7$. Partially decentralised MPC with PSF and quadratic constraint, $N = 4$, with online updating coefficient matrices for the quadratic constraint

There are also other partially decentralised control schemes for the power systems that minimise the amount of exchanged data between control areas, see, e.g. [29, 137], to eliminate a part of the installed communication links for the purpose of real-time control. Nevertheless, this chapter has offered a difference for this application in presenting a cooperative method that is useful in practice since the communication links are not assumed permanently connected. In this approach, the system is assuredly stable in a fully decentralised control architecture, yet provides a centralised MPC comparable control performance whenever the data exchanged channels resume available. The minimal off-set for the tie-line powers and system stability, which is among the main requirements, will thus be established thanks to the PSF strategy. In other words, the system operability and reliability can be guaranteed and improved by the quadratic constraint for the partially DeMPC scheme with the PSF strategy.

5.7   Concluding Remarks

This chapter has presented a partially decentralised model predictive control scheme for interconnected systems suffering from the intermittent data losses on device networks and having multiple inter-subsystem communication network topologies. The quadratic constraint is updated at each successive updating instants to deal with the data-lost process in networked control systems. The data losses are effectively compensated for in real time, right after every data loss incident, thanks to the rolling updates of the MPC that is associated a quadratic constraint on the updated control vector. A virtual perturbed cooperative-state feedback (PSF) strategy has been presented and deployed for the interconnected systems having a flexible communication network linking the neighbouring subsystems. The available state information from the connected neighbours is therefore capitalised on with this PSF strategy, yet a low computational cost of a decentralised control scheme can be maintained.

Chapter 6
Accumulative Quadratic Constraint

This chapter addresses a decentralised model predictive control (DeMPC) scheme for networks of linear systems that suffer from a coupling delay element. The accumulative quadratic supply rates are considered in this chapter for dealing with the delay in the interactive inputs of neighbouring subsystems. The stabilisability conditions are developed in conjunction with the delay robustly dissipative criteria, and in association with the newly introduced 'asymptotically accumulative quadratic constraint' (AAQC) and the 'positively accumulative quadratic constraint' (PAQC). In this chapter, the storage function of a Lyapunov–Krasovskii functional form is used in the dissipation inequality. This storage function may initially increase even when there are not any disturbances, then starts converging to a neighbourhood of the origin as a result of the employment of the AAQC-based or PAQC-based attractivity constraints in decentralised MPC. With the accumulative supply rates, the monotonically decreasing property of a single-time-step supply rate can be relaxed. The AAQC and PAQC are, thus, less conservative than the APRC and QDC in previous chapters, and well suited to the interconnected systems having a coupling delay element in this chapter.

6.1 Coupling Delay

Research on complex network systems has been increasingly attractive in various fields of science and engineering. Due to the finite speed of transmissions, traffic congestions or spreading, a time delay, called coupling delay herein, is often associated with the signals, material flows or influences travelling through a network [10, 81]. When a delay occurs in the interconnections between nodes or subsystems, as shown in Fig. 6.1, the overall performance and the system stability may be degradedly affected.

© Springer Nature Singapore Pte Ltd. 2018
A. Tri Tran C. and Q. Ha, *A Quadratic Constraint Approach to Model Predictive Control of Interconnected Systems*, Studies in Systems, Decision and Control 148, https://doi.org/10.1007/978-981-10-8409-6_6

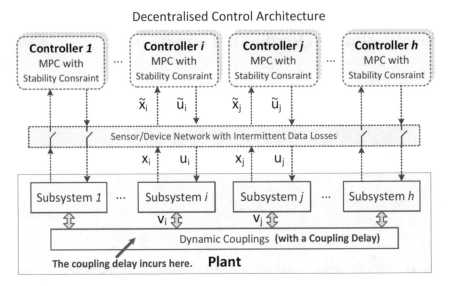

Fig. 6.1 Coupling delay in an interconnected system whose subsystems are dynamically coupled - a block diagram

Time delay systems have emerged as an important research strand [39, 48, 66, 77, 91, 94, 102, 164, 176] in the control literature with a plethora of published research papers. While laying firm foundations for the field, the past solutions to the decentralised control problems may require some artificial, or assumed, constraints on the coupling vectors and interactive variables, or on the objective function, whose presence may create some levels of conservativeness. In the presented approach with the AAQC and PAQC, the asymptotic attractivity of an interconnected system is achieved without having any artificial constraints on the coupling vectors, yet only decoupled controls are employed in the design. We only address the coupling delay of an interconnected system and a decentralised MPC scheme in this chapter.

6.2   Control System Models

Consider a network system Σ consisting of h subsystems, each denoted as $\mathscr{S}_i$, $i = 1, \ldots, h$ and has a discrete-time state-space model of the form:

$$\mathscr{S}_i : \begin{cases} x_i(k + 1) = A_i x_i(k) + B_i u_i(k) + E_i v_i(k), \\ \quad y_i(k) = C_i x_i(k), \\ \quad w_i(k) = F_i x_i(k), \end{cases} \tag{6.1}$$

$u_i \in \mathbb{U}_i \subset \mathbb{R}^{m_i}$, $y_i \in \mathbb{R}^{q_i}$, $x_i \in \mathbb{R}^{n_i}$, $v_i \in \mathbb{R}^{m_{v_i}}$, $w_i \in \mathbb{R}^{q_{w_i}}$; x_i, u_i, y_i, v_i, and w_i are, respectively, state, control, output, coupling input and output vectors. Vector d_i represents the unmeasured and bounded disturbances.

The global system Σ is then represented by a state-space model of the block-diagonal system $\mathscr{S}$ formed by h subsystems $\mathscr{S}_i$ and the interconnection process H having a coupling delay element, as follows:

$$\Sigma : \begin{cases} x(k+1) = A\,x(k) + B\,u(k) + E\,v(k), \\ \quad y(k) = C\,x(k), \\ \quad w(k) = F\,x(k), \\ \quad v(k) = H\,w(k-\tau), \end{cases} \tag{6.2}$$

where $A = \text{diag}[A_i]_1^h$, $x = [x_1^T \ldots x_h^T]^T$, and similarly for other global block-diagonal matrices B, C, E, F and the stacking vectors u, y, v, w of Σ.

The variable τ represents the number of delay time steps, $\tau \in \{1, 2, \ldots, \tau_{max}\}$, and H is the global coupling matrix whose elements are either 1 or 0 only.

From a centralised point of view, the model of Σ can be rewritten as

$$\Sigma : \begin{cases} x(k+1) = A\,x(k) + EHF\,x(k-\tau) + B\,u(k) + L\,d(k), \\ \quad y(k) = C\,x(k). \end{cases} \tag{6.3}$$

In a decentralised control scheme, the models of stand-alone subsystems are required by the local control algorithms. When all the interactive inputs and outputs vanish (i.e. $v_i = 0$ and $w_i = 0$), a stand-alone subsystem $\mathscr{S}_i$ has the state-space model of the form:

$$\mathscr{S}_i|_{\text{stand alone}} : \begin{cases} x_i(k+1) = A_i x_i(k) + B_i u_i(k), \\ \quad y_i(k) = C_i x_i(k). \end{cases} \tag{6.4}$$

The following properties are assumed in this work:

1. (A, C) and (A_i, C_i), are observable while (A, B) and (A_i, B_i) are controllable.
2. The updating time instants are synchronised between all subsystems and their local controllers.

Problem Description:

1. To implement the conventional model predictive control (MPC) algorithm [88] with the stand-alone model (6.4) in the prediction for each single subsystem $\mathscr{S}_i$ (6.1), i.e. the interactive variables v_i and w_i vanish in the objective function of each local MPC. The global system may be open-loop unstable. There are not any communication links and exchanged-information data between subsystem controllers.
2. To design an enforced attractivity constraint for each local MPC to guarantee that the plant-wide global system is asymptotically attractive. In other words, we are concerned with the design of h decoupled attractivity constraints for the local MPCs of h associated subsystems $\mathscr{S}_i$, such that the global system Σ (6.3) is

asymptotically attractive. Only local control constraints of the form $u_i(k) \in \mathbb{U}_i$, $\mathbb{U}_i$ is a compact bounded subset of $\mathbb{R}^{m_i}$, $i = 1, 2, \ldots, h$, are considered in the MPC optimisation.

3. To determine the control performance via the local MPC cost functions, which are chosen by the application on the case-by-case basis. In other words, the weighting coefficients of the MPC cost function are user-chosen parameters which are dependent upon the other performance objective required by the plant operation.

It is noted here that the coupling delay variable τ is the only new variable compared to the models in Chap. 2.

6.3 Dissipative Criteria with Accumulative Quadratic Constraints

The LMI dissipative criteria for the networks of linear systems with unknown, but bounded, coupling delay is presented in this section. In the first dissipative criterion, the *accumulatively quadratic dissipativity* (AQD) with respect to the *accumulative supply rate* (ASR) within the largest delay time interval is obtained. In the second dissipative criterion, the *delay-robust dissipativity* (DRD) for all time delay intervals that are upper bounded is acquired. The storage function of the Lyapunov–Krasovskii functional in a discrete-time form is employed in these two dissipative criteria. Both delay-dependent and delay-independent criteria are derived here.

6.3.1 Accumulatively Quadratic Dissipativity

A subsystem $\mathscr{S}_i$ (6.1) is said to be accumulatively and quadratically dissipative if it is dissipative with respect to the following accumulative supply rate:

$$\xi^{si}(\breve{y}_{si}, \breve{u}_{si}) := \breve{y}_{si}^T \breve{Q}_{si} \breve{y}_{si} + 2\breve{y}_{si}^T \breve{S}_{si} \breve{u}_{si} + \breve{u}_{si}^T \breve{R}_{si} \breve{u}_{si}, \qquad (6.5)$$

where $\breve{u}_{si}^T(k) := [\breve{v}_i^T(k) \quad \breve{u}_i^T(k) \quad \breve{d}_i^T(k)], \quad \breve{y}_{si}^T(k) := [\breve{w}_i^T(k) \quad \breve{y}_i^T(k)],$

in which $\breve{z}(k)$ is defined by $\breve{z}(k) := [z^T(k - \tau) \; z^T(k - \tau + 1) \ldots z^T(k - 1) \; z^T(k)]^T$, and

$$\breve{Q}_{si} := \begin{bmatrix} \breve{Q}_{\tau i} & 0 \\ 0 & \breve{Q}_i \end{bmatrix}, \quad \breve{R}_{si} := \begin{bmatrix} \breve{R}_{\tau i} & 0 \\ 0 & \begin{bmatrix} \breve{R}_i & 0 \\ 0 & \breve{X}_i \end{bmatrix} \end{bmatrix}, \quad \breve{S}_{si} := \begin{bmatrix} \breve{S}_{\tau i} & 0 \\ 0 & [\breve{S}_i \; 0] \end{bmatrix}.$$

All block matrices are, in turn, block-diagonal matrices of the following form:

$$\breve{M}_i := \begin{bmatrix} M_i(0) & 0 & \cdots & 0 \\ 0 & M_i(1) & \cdots & 0 \\ \vdots & \vdots & \ddots & \vdots \\ 0 & 0 & \cdots & M_i(\tau) \end{bmatrix}.$$

Now, define a delay stacking vector,

$$\breve{x}_i(k) := [x_i^T(k-\tau) \; x_i^T(k-\tau+1) \ldots x_i^T(k-1) \; x_i^T(k)]^T,$$

and assume that $x_i(j) = 0$ for all $j = -\tau, \ldots, -1, 0$, i.e. $\breve{x}_i(0) = 0$.

The following storage function of a Lyapunov–Krasovskii functional form is considered:

$$V_{si}(\breve{x}_i(k)) := x_i^T(k) P_i x_i(k) + \sum_{j=k-\tau}^{k-1} x_i^T(j) P_{\tau i}(k-j) x_i(j), \tag{6.6}$$

$$P_i = P_i^T > 0, \quad P_{\tau i}(\kappa) = P_{\tau i}^T(\kappa) > 0.$$

This Lyapunov–Krasovskii storage function $V_{si}(\breve{x}_i(k))$ can be rewritten as

$$V_{si}(\breve{x}_i(k)) = \breve{x}_i^T(k) \breve{P}_{\tau i} \breve{x}_i(k), \quad \breve{P}_{\tau i} := \begin{bmatrix} P_{\tau i}(\tau) & \cdots & 0 & 0 \\ \vdots & \ddots & \vdots & \vdots \\ 0 & \cdots & P_{\tau i}(1) & 0 \\ 0 & \cdots & 0 & P_i \end{bmatrix}. \tag{6.7}$$

The dissipation inequality corresponding to the accumulative supply rate (6.5) and the Lyapunov–Krasovskii storage function $V_{si}(\breve{x}_i(k))$ (6.6) is provided in the next definition.

Definition 6.1 A subsystem $\mathscr{S}_i$ is said to be accumulatively and quadratically dissipative (AQD) with respect to the supply rate $\xi^{si}(\breve{y}_{si}, \breve{u}_{si})$ (6.5) if there exists a non-negative C^1 function, addressed as storage function, $V_{si}(\breve{x}_i(k))$, $V_i(\breve{x}_i(0)) = 0$, such that for all $\breve{u}_{si}(k)$ and all $k \in \mathbb{Z}^+$, the following dissipation inequality is satisfied:

$$V_{si}(\breve{x}_i(k+1)) - \alpha_i \times V_{si}(\breve{x}_i(k)) \leq \xi^{si}(\breve{y}_{si}(k), \breve{u}_{si}(k)), \quad 0 < \alpha_i < 1. \tag{6.8}$$

The Lyapunov–Krasovskii storage functional of $V_{si}(\breve{x}_i(k))$ (6.6) is employed in the AQD condition. It is noted that the passivity terms describing the 'phase' property between $\breve{y}_i$ and $\breve{d}_i$ are not considered in the accumulative supply rate.

Based on this definition, we derive a delay-dependent AQD criterion in the next Theorem.

Theorem 6.1 *Consider the subsystem $\mathscr{S}_i$ (6.1) and the accumulative storage function of the Lyapunov–Krasovskii functional form (6.6). Subsystem $\mathscr{S}_i$ is*

delay-dependent AQD with respect to the accumulative supply rate $\xi^{si}(\check{y}_{si}, \check{u}_{si})$ (6.5)
if the following LMI is feasible in P_i, $P_{\tau i}(j)$, $Q_i(j)$, $S_i(j)$, $R_i(j)$, $X_i(j)$, $Q_{\tau i}(j)$,
$S_{\tau i}(j)$, $R_{\tau i}(j)$:

$$\mathcal{Q}_i + \sum_{j=1}^{\tau} \mathcal{Q}_{\tau i}(j) \prec 0, \tag{6.9}$$

$$\mathcal{Q}_i := \begin{bmatrix} \mathscr{P}_i & A_i^T P_i E_i - F_i^T S_{\tau i}(0) & A_i^T P_i B_i - C_i^T S_i(0) & A_i^T P_i L_i \\ * & E_i^T P_i E_i - R_{\tau i}(0) & E_i^T P_i B_i & E_i^T P_i L_i \\ * & * & B_i^T P_i B_i - R_i(0) & B_i^T P_i L_i \\ * & * & * & L_i^T P_i L_i - X_i(0) \end{bmatrix},$$

$$\tag{6.10}$$

where $\mathscr{P}_i := A_i^T P_i A_i - C_i^T Q_i(0)C_i - F_i^T Q_{\tau i}(0)F_i - \alpha_i \times P_i$, $0 < \alpha_i < 1$, *and*

$$\mathcal{Q}_{\tau i}(j) := \begin{bmatrix} \mathscr{P}_{\tau i}(j) & A_i^T P_{\tau i}(j)E_i - F_i^T S_{\tau i}(j) & A_i^T P_{\tau i}(j)B_i - C_i^T S_i(j) \\ * & E_i^T P_{\tau i}(j)E_i - R_{\tau i}(j) & E_i^T P_{\tau i}(j)B_i \\ * & * & B_i^T P_{\tau i}(j)B_i - R_i(j) \\ * & * & * \end{bmatrix}$$

$$\begin{matrix} A_i^T P_{\tau i}(j)L_i \\ E_i^T P_{\tau i}(j)L_i \\ B_i^T P_{\tau i}(j)L_i \\ L_i^T P_{\tau i}(j)L_i - X_i(j) \end{matrix} \Bigg], \tag{6.11}$$

in which $\mathscr{P}_{\tau i}(j) := A_i^T P_{\tau i}(j)A_i - C_i^T Q_i(j)C_i - F_i^T Q_{\tau i}(j)F_i - P_{\tau i}(j)$.

Proof By expanding the dissipation inequality of the following form

$$V_{si}(\check{x}_i(k+1)) - \alpha_i \times V_{si}(\check{x}_i(k)) \leq \xi^{si}(\check{y}_{si}(k), \check{u}_{si}(k)),$$

using the state-space model (6.1) and the accumulative storage function of the Lyapunov–Krasovskii functional form (6.6), the resultant accumulatively quadratic dissipativity criterion cast in LMI will be obtained after some matrix manipulations (see, e.g., [17, 154]). ∎

6.3.2 Delay-Robust Quadratic Dissipativity

In this subsection, a subsystem $\mathscr{S}_i$ (6.1) is said to be delay-robust dissipative, if it is quadratically dissipative with respect to the following supply rate:

$$\xi^i(y_{si}, u_{si}) = y_{si}^T Q_{si} y_{si} + 2y_{si}^T S_{si} u_{si} + u_{si}^T R_{si} u_{si}, \tag{6.12}$$

where

$$u_{si}{}^T(k) = [v_i^T(k) \quad u_i^T(k) \quad d_i^T(k)], \quad y_{si}{}^T(k) = [w_i^T(k-\tau) \quad y_i^T(k)],$$

$$Q_{si} = \begin{bmatrix} Q_{\tau i} & 0 \\ 0 & Q_i \end{bmatrix}, \quad R_{si} = \begin{bmatrix} R_{\tau i} & 0 \\ 0 & \begin{bmatrix} R_i & 0 \\ 0 & R_{di} \end{bmatrix} \end{bmatrix}, \quad S_{si} = \begin{bmatrix} S_{\tau i} & 0 \\ 0 & [S_i \ 0] \end{bmatrix}.$$

The following storage function of a Lyapunov–Krasovskii functional form will be considered in this subsection:

$$V_i(\check{x}_i(k)) = x_i^T(k)P_i x_i(k) + \sum_{j=k-\tau}^{k-1} x_i^T(j)P_{\tau i} x_i(j), \tag{6.13}$$

$$P_i = P_i^T > 0, \quad P_{\tau i} = P_{\tau i}^T > 0.$$

Only one constant $P_{\tau i}$ is considered in V_i in (6.13) instead of τ matrices $P_{\tau i}(\kappa)$ in V_{si} (6.6).

The dissipation inequality corresponding to the supply rate (6.12) and the Lyapunov–Krasovskii storage function $V_{si}(\check{x}_i(k))$ (6.13) is provided in the next definition.

Definition 6.2 A subsystem $\mathscr{S}_i$ (6.1) is said to be delay-robust dissipative (DRD) with respect to the quadratic supply rate $\xi^i(y_{si}, u_{si})$ if there exists a non-negative C^1 function, addressed as storage function, $V_i(\check{x}_i(k))$, $V_i(0) = 0$, such that for all u_{si} and all $k \in \mathbb{Z}^+$, the following dissipation inequality is satisfied:

$$V_i(\check{x}_i(k+1)) - \alpha_i \times V_i(\check{x}_i(k)) \le \xi^i(y_{si}, u_{si}), \quad 0 < \alpha_i < 1. \tag{6.14}$$

The Lyapunov–Krasovskii storage functional of $V_i(\check{x}_i(k))$ (6.13) is employed in the DRD criterion. The following theorem provides a delay-independent DRD criterion.

Theorem 6.2 *Consider the subsystem $\mathscr{S}_i$ (6.1) and the storage function of the Lyapunov–Krasovskii functional form (6.13). Subsystem $\mathscr{S}_i$ is delay-independent delay-robust dissipative (DRD) with respect to the supply rate $\xi^i(y_{si}, u_{si})$ (6.12), if the following LMI is feasible in P_i, $P_{\tau i}$, Q_i, S_i, R_i, X_i, $Q_{\tau i}$, $S_{\tau i}$, $R_{\tau i}$:*

$$\begin{bmatrix} \mathscr{P}_i & 0 & A_i^T P_i E_i & A_i^T P_i B_i - C_i^T S_i & A_i^T P_i L_i \\ * & -\alpha_i \times P_{\tau i} - F_i^T Q_{\tau i} F_i & -F_i^T S_{\tau i} & 0 & 0 \\ * & * & E_i^T P_i E_i - R_{\tau i} & E_i^T P_i B_i & E_i^T P_i L_i \\ * & * & * & B_i^T P_i B_i - R_i & B_i^T P_i L_i \\ * & * & * & * & L_i^T P_i L_i - X_i \end{bmatrix} \preccurlyeq 0, \tag{6.15}$$

where $\mathscr{P}_i = A_i^T P_i A_i - \alpha_i \times P_i + P_{\tau i} - C_i^T Q_i C_i$, $0 < \alpha_i < 1$.

Proof By expanding the dissipation inequality of the following form

$$V_i(\check{x}_i(k+1)) - \alpha_i V_i(\check{x}_i(k)) \le \xi^i(y_{si}(k), u_{si}(k)),$$

using the state-space model (6.1) and the storage function of the Lyapunov–Krasovskii functional form (6.13), the resultant dissipativity criterion cast in LMI in (6.15) will be obtained after some matrix manipulations. ∎

The following result provides a delay-dependent delay-robust dissipativity criterion for $\mathscr{S}_i$. The delay-dependent criterion is an extension of the delay-independent criterion in Theorem 6.2.

Theorem 6.3 *Consider the subsystem $\mathscr{S}_i$ (6.1) and the storage function of the Lyapunov–Krasovskii functional form (6.13). Subsystem $\mathscr{S}_i$ is delay-dependent delay-robust dissipative (DRD) with respect to the supply rate $\xi^i(y_{si}, u_{si})$ (6.12), if the following LMI is feasible in P_i, $P_{\tau i}$, Q_i, S_i, R_i, X_i, $Q_{\tau i}$, $S_{\tau i}$, $R_{\tau i}$:*

$$\begin{bmatrix} \mathscr{P}_{\tau i} & A_i^{\tau T} \mathscr{P}_i \mathscr{B}_i & A_i^{\tau+1^T} P_i E_i - F_i^T S_{\tau i} & A_i^{\tau T} (A_i^T P_i B_i - C_i^T S_i) & A_i^{\tau+1^T} P_i L_i \\ * & \mathscr{B}_i^T \mathscr{P}_i \mathscr{B}_i & \mathscr{B}_i^T A_i^T P_i E_i & \mathscr{B}_i^T (A_i^T P_i B_i - C_i^T S_i) & \mathscr{B}_i^T A_i^T P_i L_i \\ * & * & E_i^T P_i E_i - R_{\tau i} & E_i^T P_i B_i & E_i^T P_i L_i \\ * & * & * & B_i^T P_i B_i - R_i & B_i^T P_i L_i \\ * & * & * & * & L_i^T P_i L_i - X_i \end{bmatrix} \prec 0,$$
$$\tau = 1 \ldots \tau_{max},$$
(6.16)

where
$$\mathscr{P}_i = A_i^T P_i A_i - \alpha_i \times P_i + P_{\tau i} - C_i^T Q_i C_i,$$

$$\mathscr{P}_{\tau i} = A_i^{\tau T} \mathscr{P}_i A_i^\tau - \alpha_i \times P_{\tau i} - F_i^T Q_{\tau i} F_i,$$

$$\mathscr{B}_i = [A_i^{\tau-1} B_{si} \ \ldots \ A_i B_{si} \ B_{si}].$$

Proof From the state-space model (6.1), we obtain $x_i = A_i^\tau x_i(-\tau) + \mathscr{B}_i \check{u}_{\tau i}$, where

$$\mathscr{B}_i = [A_i^{\tau-1} B_{si} \ \ldots \ A_i B_{si} \ B_{si}], \quad B_{si} = [B_i \ \ E_i \ \ L_i],$$
$$\check{u}_{\tau i} = [u_{si}(k-\tau) \ \ldots \ u_{si}(k-1)], \quad u_{si} = [u_i \ \ d_i \ \ v_i].$$

Substituting to the dissipation inequality (6.14) yields

$$(*)\begin{bmatrix} A_i^{\tau T} \mathscr{P}_i A_i^\tau - \alpha_i P_{\tau i} - F_i^T Q_{\tau i} F_i & A_i^{\tau T} \mathscr{P}_i \mathscr{B}_i & A_i^{\tau+1^T} P_i E_i - F_i^T S_{\tau i} \\ * & \mathscr{B}_i^T \mathscr{P}_i \mathscr{B}_i & \mathscr{B}_i^T A_i^T P_i E_i \\ * & * & E_i^T P_i E_i - R_i \\ * & * & * \\ * & * & * \end{bmatrix}$$

$$\begin{bmatrix} A_i^{\tau T}(A_i^T P_i B_i - C_i^T S_i) & A_i^{\tau+1^T} P_i L_i \\ \mathscr{B}_i^T(A_i^T P_i B_i - C_i^T S_i) & \mathscr{B}_i^T A_i^T P_i L_i \\ E_i^T P_i B_i & E_i^T P_i L_i \\ B_i^T P_i B_i - R_i & B_i^T P_i L_i \\ * & L_i^T P_i L_i - R_{di} \end{bmatrix} \begin{bmatrix} x_i(-\tau) \\ \breve{u}_{\tau i} \\ v_i \\ u_i \\ d_i \end{bmatrix} \preccurlyeq 0,$$

which is fulfilled by (6.16). ∎

The dissipative criteria in this section will be used in the stabilisability conditions in the next section.

6.4 Accumulative Constraints and Stabilisability Conditions

6.4.1 Positive Realness Constraint and Summability with Time Delay

Consider the quadratic function ξ_i with respect to the output and input pair of (y_i, u_i),

$$\xi_i(y_i, u_i) := u_i^T Q_{ic} u_i + 2u_i^T S_{ic}^T y_i + y_i^T R_{ic} y_i. \tag{6.17}$$

(1) The $\tau-$ positive realness constraint, or τ–PRC is defined as follows: The nominal subsystem $\mathscr{S}_i$ (6.1) with $d_i(k) = 0$ is then said to be bounded by the τ–PRC with respect to $\xi_i(y_i, u_i)$ if the accumulation of $\xi_i(y_i, u_i)$ over the time interval $(k-\tau, k)$ along the trajectory, denoted as $\xi_{ia}(k, \tau)$, is non-negative; i.e.

$$\xi_{ia}(k, \tau) := \sum_{\kappa=k-\tau}^{k} \xi_i(y_i(\kappa), u_i(\kappa)) \geqslant 0, \quad \text{for } \tau \in \{1, 2, \dots, \tau_{\max}\}. \tag{6.18}$$

The matrices Q_{ic}, S_{ic}, R_{ic} will be determined in alliance with the LMI dissipative criteria in the previous section.

(2) The positive summability is defined as follows: The nominal subsystem $\mathscr{S}_i$ (6.1) with $d_i(k) = 0$ is said to be positively summable with respect to $\xi_i(y_i, u_i)$ if the accumulation of $\xi_i(., .)$ along the trajectory, denoted as $\xi_{is}(k)$, is non-negative; i.e.

$$\xi_{is}(k) := \sum_{\kappa=0}^{k} \xi_i(y_i(\kappa), u_i(\kappa)) \geqslant 0. \tag{6.19}$$

The two asymptotic constraints are first defined in the next subsection before the respective stabilisability conditions are derived correspondingly.

6.4.2 Accumulative Quadratic Constraint

The τ–PRC is achievable with the asymptotically accumulative quadratic constraint (AAQC) defined in the following:

Definition 6.3 The controlled motion $(y_i(k), u_i(k))$ of $\mathscr{S}_i$ (6.1) is said to satisfy the asymptotically accumulative quadratic constraint, practically, with respect to ξ_i and the coupling delay τ, or simply AAQC, if there is $\gamma_{ia} \in \mathbb{R}_+$, $\gamma_{ia} < 1$, and a small number $\varepsilon_{ia} < 0$, such that the following condition holds for all $k > \tau_{\max}$:

$$\begin{cases} \xi_{ia}(k, \tau) \geqslant \gamma_{ia} \times \xi_{ia}(k-1, \tau) + \varepsilon_{ia}, & \text{if } \xi_{ia}(k-1, \tau) < 0, \\ \xi_{ia}(k, \tau) \geqslant \varepsilon_{ia}, & \text{if } \xi_{ia}(k-1, \tau) \geqslant 0. \end{cases} \quad (6.20)$$

The positive summability is achievable with the positively accumulative quadratic constraint (PAQC) defined in the following:

Definition 6.4 The controlled motion $(y_i(k), u_i(k))$ of $\mathscr{S}_i$ (6.1) is said to satisfy the positively accumulative quadratic constraint, practically, with respect to ξ_i, or simply PAQC, if there is $\gamma_{is} \in \mathbb{R}_+$, $\gamma_{is} < 1$, and a small number $\varepsilon_{is} < 0$, such that the following conditions hold for all $k > \tau_{\max}$:

$$\xi_{is}(k) \geqslant \varepsilon_{is} \quad \text{and} \quad \begin{cases} \xi_{is}(k+1) < +\infty, & \text{if } \xi_i(k+1) \geqslant 0, \\ \xi_{is}(k+1) \leqslant \gamma_{is} \times \xi_{is}(k) + \varepsilon_{is}, & \text{if } \xi_i(k+1) < 0. \end{cases} \quad (6.21)$$

The two stability conditions corresponding to the two accumulative constraints, AAQC and PAQC, are stated in the next subsections.

6.4.3 Input-to-Power-and-State Stabilisation

The input-to-power-and-state stabilisability (IpSS) is firstly recalled here.

Consider Σ (6.3) with $\tau = 0$ and $C = I_n$ (i.e. $y = x$), and the real-valued supply-rate function $\xi(x, u) : \mathbb{R}^n \times \mathbb{R}^m \to \mathbb{R}$, which can be piece-wise continuous, or continuous, in x and u.

The system Σ is said to be IpSS stabilisable if there are two functions α_i of class $\mathscr{K}\mathscr{L}$, $i = \{0, 1\}$, a finite initial supply rate $\xi(k_0)$ and a function γ of class $\mathscr{K}$, such that for each bounded $d(k)$ and each initial state $x(k_0)$, the following inequality is satisfied for all $k > k_0$:

$$\|x(k, x(k_0), d)\| \leqslant \alpha_0(\|x(k_0)\|, k - k_0) + \alpha_1(|\xi(k_0)|, k - k_0) + \gamma(\|d\|_\infty),$$
$$(6.22)$$

with some admissible control sequence $\{u(k) \in \mathbb{U}\}$, where $\xi(k_0) := \xi\big(x(k_0), u(k_0)\big)$.

The IpSS for interconnected systems having coupling delays is defined next.

Definition 6.5 The controlled system Σ (6.3) is said to be IpSS stabilisable with the AAQC if there are two functions α_i of class $\mathcal{K}\mathcal{L}$, $i \in \{0, 1\}$, a finite initial supply rate $\xi_{(k_0)}$ and a function γ of class $\mathcal{K}$, such that for each bounded $d(k)$ and each initial state $x(k_0) = x_{k_0}$, the following inequality is satisfied for all $k > k_0 > \tau_{\max}$:

$$\|x(k, x_{k_0}, d)\| \leqslant \alpha_0(\|x_{k_0}\|, k - k_0) + \alpha_1(|\xi_{ia}(k_0)|, k - k_0) + \gamma(\|d\|_\infty), \quad (6.23)$$

with some admissible control sequences $\{u(k) \in \mathbb{U}\}$.

Definition 6.6 The controlled system Σ (6.3) is said to be IpSS stabilisable with the PAQC if there are two functions α_i of class $\mathcal{K}\mathcal{L}$, $i \in \{0, 1\}$, a finite initial supply rate $\xi_{(k_0)}$ and a function γ of class $\mathcal{K}$, such that for each bounded $d(k)$ and each initial state $x(k_0) = x_{k_0}$, the following inequality is satisfied for all $k > k_0 > \tau_{\max}$:

$$\|x(k, x_{k_0}, d)\| \leqslant \alpha_0(\|x_{k_0}\|, k - k_0) + \alpha_1(\xi_{is}(k_0), k - k_0) + \gamma(\|d\|_\infty), \quad (6.24)$$

with some admissible control sequences $\{u(k) \in \mathbb{U}\}$.

6.4.4 AAQC-Based Stabilisability Condition

To obtain the AAQC-based stabilisability condition for interconnected systems having coupling delays, we denote the following:

$$\xi_{ia}(k, \tau) := \breve{u}_i^T \breve{Q}_{ic} \breve{u}_i + 2\breve{y}_i^T \breve{S}_{ic} \breve{u}_i + \breve{y}_i^T \breve{R}_{ic} \breve{y}_i,$$

where $\quad \breve{Q}_{ic} := \mathrm{diag}[Q_{ic}]_1^{\tau+1}$, $\breve{S}_{ic} := \mathrm{diag}[S_{ic}]_1^{\tau+1}$, $\breve{R}_{ic} := \mathrm{diag}[R_{ic}]_1^{\tau+1}$,

and $\breve{u}_i$, $\breve{y}_i$ are the stacking vectors of u_i and y_i, respectively.

The LMI-based attractivity condition is established from the accumulative dissipativity of all subsystems and their AAQCs, as stated in Theorem 6.4 which provides delay-dependent conditions.

Theorem 6.4 *Consider h subsystems $\mathscr{S}_i$ (6.1), $i = 1, 2, \ldots, h$, interconnected by the global coupling process H having the coupling day element τ, and h respective non-negative storage functions of the Lyapunov–Krasovskii functional form (6.6). The controlled system Σ (6.3) is asymptotically attractive when $d_i(k) \equiv 0$, and IpSS stabilised with the AAQC when $d_i(k) \neq 0$, if the following LMIs are feasible in $\breve{Q}_i, \breve{S}_i, \breve{R}_i, \breve{X}_i, \breve{Q}_{ic}, \breve{S}_{ic}, \breve{R}_{ic}, \breve{Q}_{\tau i}, \breve{S}_{\tau i}, \breve{R}_{\tau i}$:*

$$\text{diag}[\check{Q}_{\tau i}]_1^h + \check{H}^T\text{diag}[\check{R}_{\tau i}]_1^h\check{H} + \check{H}^T\text{diag}[\check{S}_{\tau i}^T]_1^h + \text{diag}[\check{S}_{\tau i}]_1^h\check{H} \prec 0, \quad (6.25)$$

$$\begin{bmatrix} \check{Q}_i + \check{R}_{ic} & \check{S}_i + \check{S}_{ic} \\ * & \check{R}_i + \check{Q}_{ic} \end{bmatrix} \prec 0, \quad (6.26)$$

and (6.9), for i = 1 ... h,

where $\check{H} := \text{diag}\{H\}_{\tau_{\max}}$, and the AAQC (6.20) is feasible in $u_i(k)$ for all $i \in \{1, 2, \ldots, h\}$.

Proof This is a direct result of Theorem 2.1 with the state vector $\check{x}_{si}$, control vector $\check{u}_{si}$ and output vector $\check{y}_{si}$ for the case of $d(k) = 0$. The proof for the case of $d(k) \neq 0$ is similar to that in [161] ∎

6.4.5 PAQC-Based Stabilisability Condition

By defining the global coefficient matrices

$$Q = \text{diag}[Q_i]_1^h, \; S = \text{diag}[S_i]_1^h, \; R = \text{diag}[R_i]_1^h, \; R_d = \text{diag}[R_{id}]_1^h,$$

$$Q_\tau = \text{diag}[Q_{\tau i}]_1^h, \; S_\tau = \text{diag}[S_{\tau i}]_1^h, \; R_\tau = \text{diag}[R_{\tau i}]_1^h, \; R_c = \text{diag}[R_{ic}]_1^h,$$

$$S_c = \text{diag}[S_{ic}]_1^h, \; Q_c = \text{diag}[Q_{ic}]_1^h, \; Q_d = \text{diag}[Q_{id}]_1^h,$$

the stabilisability conditions can be stated in Theorems 6.5 and 6.6 below. The LMI-based stability condition is derived from the delay-robust dissipativity of subsystems and their PAQCs. The following storage function of a Lyapunov–Krasovskii functional form is employed in these two theorems:

$$V_{si}(k) = x_i^T(k)P_ix_i(k) + \sum_{j=k-\tau}^{k-1} x_i^T(j)P_{\tau i}x_i(j), \quad P_i \succ 0, \; P_{\tau i} \succ 0. \quad (6.27)$$

Denote $\xi_{\Sigma\Sigma c}(\kappa) = \sum_h \xi_{is}(\kappa)$, and assume that $x_i(k) = 0$, $i = 1, 2, \ldots, h$, in the time intervals prior to 0: $-\tau_{\max} \leqslant k \leqslant 0$ and $\xi_{\Sigma\Sigma c}(-1) > 0$.

Theorem 6.5 *Let $0 < \alpha < 1$. Consider h subsystems $\mathscr{S}_i$ (6.1), $i = 1, 2, \ldots, h$, interconnected by the global coupling process H having the coupling day element τ, and h respective non-negative storage functions of the Lyapunov–Krasovskii functional form (6.13). The controlled system Σ (6.3) is asymptotically attractive when $d_i(k) \equiv 0$, and IpSS stabilised with the PAQC when $d_i(k) \neq 0$, if the following LMIs are feasible in P_i, $P_{\tau i}$, Q_i, S_i, R_i, X_i, $Q_{\tau i}$, $S_{\tau i}$, $R_{\tau i}$, Q_{ic}, S_{ic}, R_{ic}, Z_{ic}:*

$$F^T Q_\tau F + F^T S_\tau^T H F + F^T H^T S_\tau F + F^T H^T R_\tau H F \prec 0,$$

$$\begin{bmatrix} Q_i + \frac{\alpha}{2} R_{ic} & S_i + \frac{\alpha}{2} S_{ic} \\ * & R_i + \frac{\alpha}{2} Q_{ic} \end{bmatrix} \prec 0, \tag{6.28}$$

and (6.15), for all $i \in \{1, 2, \ldots h\}$

and the PAQC (6.21) is feasible in $u_i(k)$ for all $i \in \{1, 2, \ldots h\}$.

Proof Firstly, rewrite $\xi_s(y(k), u(k)) = \sum_{i=1}^{h} \xi_{is}(y_i(k), u_i(k))$, as

$$\xi_s(y(k), u(k)) = y^T(k) R_c y(k) + 2y^T(k) S_c u(k) + u^T(k) Q_c u(k).$$

Then, consider the large-scale vectors x, u, d, y, v and w being integrated with the coupling delay process H in a supply rate of the form

$$\xi(w(k), y(k); v(k), u(k), d(k))$$
$$= y^T(k) Q y(k) + 2y^T(k) S u(k) + u^T(k) R u(k) + d^T(k) X d(k)$$
$$+ w^T(-\tau)(k) Q_\tau w(-\tau)(k) + 2w^T(-\tau)(k) S_\tau H w(-\tau)(k)$$
$$+ w^T(-\tau)(k) H^T R_\tau H w(-\tau)(k).$$

The weighted sum of the two above quadratic supply rates, denoted as $\sigma(k)$, is obtained in the following after omitting the time index k for conciseness:

$$\sigma = \xi(w, y; v, u, d) + \frac{\alpha}{2} \xi^c(u, d; y) \tag{6.29}$$

$$= y^T (Q + \frac{\alpha}{2} R_c) y + 2y^T (S + \frac{\alpha}{2} S_c) u + u^T (R + \frac{\alpha}{2} Q_c) u$$
$$+ d^T X d + x^T(-\tau) F^T Q_\tau F x(-\tau) + 2x^T(-\tau) F^T S_\tau H F x(-\tau)$$
$$+ x^T(-\tau) F^T H^T R_\tau H F x(-\tau),$$
$$= x^T(-\tau)(F^T Q_\tau F + 2F^T S_\tau H F + F^T H^T R_\tau H F) x(-\tau) \tag{6.30}$$
$$+ \begin{bmatrix} y \\ u \end{bmatrix}^T \begin{bmatrix} Q + \frac{\alpha}{2} R_c & S + \frac{\alpha}{2} S_c \\ * & R + \frac{\alpha}{2} Q_c \end{bmatrix} \begin{bmatrix} y \\ u \end{bmatrix} + d^T X d.$$

Due to (6.28), we have $\sigma(k) < 0$ for all $[x^T(k-\tau) \ y^T(k) \ u^T(k) \ d^T(k)]^T \neq 0$. As $\xi_{\Sigma\Sigma c}(\kappa) = \sum_h \xi_{is}(\kappa)$, we also have $\xi_{\Sigma\Sigma c}(\kappa) := \sum_{k=0}^{\kappa} \xi_s(y_i(k), u_i(k))$, and thus,

$$\xi_{\Sigma\Sigma c}(\kappa) \geqslant 0 \text{ and } \xi_{\Sigma\Sigma c}(\kappa) \leqslant \gamma \xi_{\Sigma\Sigma c}(\kappa) \text{ for } \xi_{is}(k) < 0, \text{ i.e. } \xi_{is} = 0$$

according to inequalities in $\xi_{is}(\kappa)$ for $\gamma := \max_{i=1\ldots h} \gamma_i$.

Combining the above result with the dissipation inequality (6.14), which is fulfilled by (6.15), and choosing $\gamma \equiv \alpha \, \forall k < +\infty$ and $\alpha \equiv \max_{i=1...h} \alpha_i$, $i = 1 \ldots h$, we obtain for

$$V_\Sigma(x(k)) := x^T(k)Px(k) + \sum_{j=k-\tau}^{k-1} x^T(j)P_\tau x(j), \ P = \text{diag}[P_i]_1^h, \ P_\tau = \text{diag}[P_{\tau i}]_1^h,$$

the following inequality:

$$[V_\Sigma(x(k+1)) + \xi_{\Sigma\Sigma c}(k+1)] - [\alpha \, V_\Sigma(x(k)) + \frac{\alpha}{2}\xi_{\Sigma\Sigma c}(k)] \leqslant \sigma(k) + \frac{\alpha}{2}\xi_{\Sigma\Sigma c}(k-1), \tag{6.31}$$

if $d(k) = 0$.

From $\sigma(k) < 0, 0 < \alpha < 1, V_\Sigma(x(k)) \geqslant 0$ and $\xi_{\Sigma\Sigma c}(k) \geqslant 0 \, \forall k \in \mathbb{Z}^+$, it is implied by (6.31) that

$$[V_\Sigma(x(k+1)) + \xi_{\Sigma\Sigma c}(k+1)] \leqslant \alpha \, [V_\Sigma(x(k)) + \xi_{\Sigma\Sigma c}(k)] + \frac{\alpha}{2}\xi_{\Sigma\Sigma c}(k-1) \, \forall k \in \mathbb{Z}^+.$$

Now, denote $W_\Sigma(k) := V_\Sigma(x(k)) + \xi_{\Sigma\Sigma c}(k)$. With $\xi_{\Sigma\Sigma c}(k) \leqslant \alpha \, \xi_{\Sigma\Sigma c}(k-1)$ (6.21) and $0 < \alpha < 1$, by induction we have

$$W_\Sigma(k) \leqslant \alpha^{k-k_0} W_\Sigma(k_0) + \alpha^{k-k_0}\frac{k-k_0}{2}\xi_{\Sigma\Sigma c}(k_0-1).$$

We then need to prove that $W_\Sigma(k) \to 0$ as $k \to +\infty$, i.e. for each $\beta > 0$, there is $0 < k(\beta) < +\infty$ such that $W_\Sigma(k) \leqslant \beta \, \forall k \geqslant k(\beta)$, for a chosen $\alpha < 1$ such that $\alpha^{k-k_0}\frac{k-k_0}{2} \to 0$ as $k \to +\infty$. Indeed, for each $\beta \geqslant 0$, there exist two time instants $\bar{k}$ and $\tilde{k}$ such that

$$\alpha^{k-k_0}\frac{k-k_0}{2}\xi_{\Sigma\Sigma c}(k_0-1) \leqslant \frac{\beta}{2} \, \forall k > \bar{k} \text{ and } \alpha^{k-\tilde{k}}W_\Sigma(k) \leqslant \frac{\beta}{2} \, \forall k > \tilde{k}.$$

Therefore, there exists $\check{k} = \sup(\bar{k}, \tilde{k})$ such that for each $\beta \geqslant 0$,

$$W_\Sigma(k) \leqslant \frac{\beta}{2} + \frac{\beta}{2} = \beta \, \forall k \geqslant \check{k}.$$

Since Σ is observable, $\xi_{\Sigma\Sigma c}(k) \geqslant 0 \, \forall k \in \mathbb{Z}^+$, $P_i \succ 0$ and $P_{\tau i} \succ 0$, it obvious that $\xi_{\Sigma\Sigma c}(k) \to 0$ as $k \to +\infty$ and $V_\Sigma(x(k)) \to 0$ as $k \to +\infty$, which implies $\|x(k)\| \to 0$ as $k \to +\infty$. The convergence of the global state vector when $d(k) = 0$ is thus obtained.

Then, similarly to the proof of Theorem 2 in [161], the IpSS is obtained with the PAQC. The proof is complete ∎

On the ground of Theorem 6.5, the delay-dependent stability condition is stated in Theorem 6.6 below.

Theorem 6.6 *Let* $0 < \alpha < 1$. *Consider* h *subsystems* $\mathscr{S}_i$ (6.1), $i = 1, 2, \ldots, h$, *interconnected by the global coupling process* H *having the coupling day element* τ, *and* h *respective non-negative storage functions of the Lyapunov–Krasovskii functional form* (6.13). *The controlled system* Σ (6.3) *is asymptotically attractive when* $d_i(k) \equiv 0$, *and IpSS stabilised when* $d_i(k) \neq 0$, *if the PAQCs* (6.21) *and the following LMIs are feasible in* P_i, $P_{\tau i}$, Q_i, S_i, R_i, X_i, $Q_{\tau i}$, $S_{\tau i}$, $R_{\tau i}$, Q_{ic}, S_{ic}, R_{ic}, *by the block-diagonal matrices* P, P_τ, Q, S, R, X, Q_τ, S_τ, R_τ, Q_c, S_c, R_c:

$$
\begin{bmatrix}
\left\{ \begin{matrix} (*)(\frac{\alpha}{2} R_c + Q)C(A)^\lambda \\ +\mathscr{H} \end{matrix} \right\} & (A^T)^\lambda C^T (\frac{\alpha}{2} R_c + Q) & (A^T)^\lambda C^T (\frac{\alpha}{2} S_c + S) \\
* & \frac{\alpha}{2} R_c + Q & \frac{\alpha}{2} S_c + S \\
* & * & \frac{\alpha}{2} Q_c + R
\end{bmatrix} \prec 0,
$$

$$
and\ (6.16)\ for\ i = 1, 2, \ldots, h, \tag{6.32}
$$

where $\mathscr{H} := F^T Q_\tau F + F^T S_\tau^T H F + F^T H^T S_\tau F + F^T H^T R_\tau H F$ and $\lambda = \tau_{\max}$.

Proof From the definition of $\sigma(k)$ in the proof of Theorem 6.5 and its expansion, by rewriting the state vector as $x(k) = (A)^\tau x(k - \tau) + \mathscr{B} \breve{u}_\tau$, where $\breve{u}_\tau$ is a vector sequence formed by stacking h vectors $\breve{u}_{\tau i}$, and $\mathscr{B} := \text{diag}[\mathscr{B}_i]_1^h$, we obtain

$$
\sigma = [x^T(-\tau)(A^T)^\tau + \breve{u}_\tau^T \mathscr{B}^T]C^T(\frac{\alpha}{2} R_c + Q)C[(A)^\tau x(-\tau) + \mathscr{B} \breve{u}_\tau]
$$
$$
+ 2[x^T(-\tau)(A^T)^\tau + \breve{u}_\tau^T \mathscr{B}^T]C^T(\frac{\alpha}{2} S_c + S)u + u^T(\frac{\alpha}{2} Q_c + R)u + d^T X d
$$
$$
+ x^T(-\tau)(F^T Q_\tau F + F^T S_\tau^T H F + F^T H^T S_\tau F + F^T H^T R_\tau H F)x(-\tau).
$$

Then, due to (6.32) with $\tau \equiv \lambda$, we have $\sigma(k) < 0$ for all $[x^T(k - \tau)\ \breve{u}_\tau^T(k)\mathscr{B}^T C^T\ u^T(k)]^T \neq 0$. Similarly to the proof of Theorem 6.5, the attractivity for the maximum delay interval $\tau_{\max}$ when $d(k) = 0$, and similarly to the proof of Theorem 2 in [161], the IpSS with the PAQC when $d(k) \neq 0$, are obtained. ∎

The AAQC and PAQC will be employed in a decentralised MPC (DeMPC) problem to be stated in the next section.

6.5   Decentralised Model Predictive Control

To streamline the presentation, the model predictive control optimisation is re-formulated in this section. In a model predictive control problem, the two main elements of interest are the predictive control sequence and initial state vector. The predictive control sequence $\{u_i(k), u_i(k+1), \ldots, u_i(k+N)\}$ is denoted by $\hat{u}_i(k, x_i^k, N)$ or simply $\hat{u}_i$, where N is the predictive horizon, $x_i(k)$ is a known state vector and k is the current time step in the discrete-time domain. The predictive state sequence, denoted as $\hat{x}_i(k, \hat{u}_i, N) = \left\{x_i^{\hat{u}_i}(k+1, x_i(k)), \ldots, x_i^{\hat{u}_i}(k+N, x_i(k))\right\}$, is obtained as a result of applying the predictive control sequence $\hat{u}$ to the open-loop subsystem $\mathscr{S}_i|_{\text{stand alone}}$ (6.4). The control objective is to steer the state to the origin. For a known current-time state vector $x_i(k)$, the cost function is generally defined by

$$\mathscr{J}^i\left(\hat{u}_i(k, x_i^k, N)\right) = F_i\left(x_i(k+N)\right) + \sum_{j=k}^{k+N} \ell_i\left(x_i(j), u_i(j)\right).$$

The terminal time $k+N$, which could vary with time k but should not be decreasing, is often referred to as a predictive horizon. $F_i\left(x_i(k+N)\right)$ is the terminal cost function which is used to emulate the infinite-horizon optimising performance for the above finite-horizon problem. At every time step k, the open-loop optimisation problem of minimising $\mathscr{J}^i\left(\hat{u}_i(k, x_i^k, N)\right)$ subject to $u_i(j) \in \mathbb{U}_i$ is solved to yield the optimizing solution of the form $\hat{u}_i^0(k, x_i^k, N) = \{u_i^0(k, x_i(k)), u_i^0(k+1, x_i(k)), \ldots, u_i^0(k+N, x_i(k))\}$. The first control move $u_i^0(k, x_i(k))$ is then applied to control the plant. The whole process is repeated in a rolling or receding manner in the next iteration. Since N is finite, a global minimum of $\mathscr{J}^i\left(\hat{u}_i(k, x_i^k, N)\right)$ exists if $\ell_i(x_i, u_i)$ are continuously differentiable, $\mathbb{U}_i$ is bounded.

Here, the MPC problem is formulated with a new attractivity constraint developed from the AAQC.

6.5.1   Objective Function

For each stand-alone subsystem or subsystem in a decentralised control scheme, a quadratic objective function of the states and control inputs is considered in association with the adequately chosen weighting matrices $\mathscr{Q}^x$, $\mathscr{Q}^u$, as follows:

$$\mathscr{J}^i\left(\hat{u}_i(k, x_i^k, N)\right) = \sum_{j=1}^{N+1} x_i^T(k+j)\mathscr{Q}_i^x x_i(k+j) + \sum_{j=0}^{N} u_i^T(k+j)\mathscr{Q}_i^u u_i(k+j),$$

(6.33)

which is subject to the equality constraint of the state-space model (6.1) and the inequality constraint deduced from physical limits of x_i and u_i.

The problem is formulated by predicting the state vector x_i in N time steps using its current value. The predictive state vector is obtained from the perfect model without the input disturbance vector d_i, as follows:

$$\hat{x}_i(k, \hat{u}_i, N) = \Theta_i x_i(k) + \Gamma_i \hat{u}_i(k, x_i^k, N),$$

where $\hat{u}_i(k, x_i^k, N) := [u_i(k)\, u_i(k+1) \ldots u_i(k+N)]^T$, $\hat{x}_i(k, \hat{u}_i, N) := [x_i(k+1)$ $x_i(k+2) \ldots x_i(k+N+1)]^T$,

$$\Gamma_i := \begin{bmatrix} B_i & 0 & \ldots & 0 \\ A_i B_i & B_i & \ldots & 0 \\ \ldots & \ldots & \ldots & \ldots \\ A_i^N B_i & \ldots & A_i B_i & B_i \end{bmatrix}, \quad \text{and} \quad \Theta_i := \begin{bmatrix} A_i \\ A_i^2 \\ \ldots \\ A_i^{N+1} \end{bmatrix}.$$

By denoting $x_i(k, \hat{u}_i, N)$ as $\hat{x}_i(k)$ or just $\hat{x}_i$, and $u_i(k, x_i^k, N)$ as $\hat{u}_i(k)$ or just $\hat{u}_i$, we have:

$$\hat{x}_i = r_i^k + \Gamma_i \hat{u}_i, \tag{6.34}$$

where $r_i^k := \Theta_i x_i(k)$.

The objective function is now rewritten as,

$$\mathscr{J}^i \left(\hat{u}_i(k, x_i^k, N) \right) = [r_i^k + \Gamma_i \hat{u}_i]^T \mathscr{Q}_{iN}^x [r_i^k + \Gamma_i \hat{u}_i] + \hat{u}_i^T \mathscr{Q}_{iN}^u \hat{u}_i,$$

where $\mathscr{Q}_{iN}^x := \text{diag}[\mathscr{Q}_i^x]_1^N$, $\mathscr{Q}_{iN}^u := \text{diag}[\mathscr{Q}_i^u]_1^N$. It can be cast in the form of

$$\mathscr{J}^i \left(\hat{u}_i(k, x_i^k, N) \right) = \hat{u}_i^T \Phi_i \hat{u}_i + 2 \Upsilon_i^k \hat{u}_i + \delta_i^k, \tag{6.35}$$

where $\Phi_i := \Gamma_i^T \mathscr{Q}_{iN}^x \Gamma_i + \mathscr{Q}_{iN}^u$, $\Upsilon_i^k := r_i^{k^T} \mathscr{Q}_{iN}^x \Gamma_i$, $\delta_i^k := r_i^{k^T} \mathscr{Q}_{iN}^x r_i^k$.

6.5.2 Constraint on Decision Variables

Let us consider a bounded constraint of the following form:

$$\|u_i\|_2^2 \leqslant \eta_i. \tag{6.36}$$

The constraint set for the variable vector $\hat{u}_i$ will be as follows:

$$\hat{\mathbb{U}}_i := \{\hat{u}_i = [u_i^T(k) \ldots u_i^T(k+N_i)]^T : \|u_i(.)\|_2^2 \leqslant \eta_i\}. \tag{6.37}$$

6.5.3   Semi-definite Programming

The problem of minimising $\mathscr{J}^i\left(\hat{u}_i(k, x_i^k, N)\right)$ subject to the required constraints can be written for semi-definite programming (SDP) by applying the Schur complement to (6.35) as:

$$\min_{\hat{u}_i,\, \gamma_i(k)} \gamma_i(k) \tag{6.38}$$

$$\text{subject to} \quad \gamma_i(k) > 0,$$

$$\begin{bmatrix} -\gamma_i(k) + \delta_i^k + 2\Upsilon_i^k \hat{u}_i & \hat{u}_i^T \\ * & -\Phi_i^{-1} \end{bmatrix} \preccurlyeq 0, \quad \text{and} \quad (6.37).$$

The parameters Φ_i, Υ_i^0 and δ_i^0 are found from the state-space realisation matrices A_i, B_i, the weighting matrices $\mathscr{Q}_{iN}^x$, $\mathscr{Q}_{iN}^u$ of the cost function. Υ_i^k and δ_i^k are subsequently updated at every time step using the updated state vector $x_i(k)$. For each subsystem, the optimisation problem (6.38) is solved by the local controller for the minimising vector sequence $\hat{u}_i$ which comprises N elements of u_i. Only the first element $\mathring{u}_i(k)$ of the sequence $\hat{u}_i$ is applied to control the plant. This rolling process is repeated at the next time step and continues thereon. Solving this problem does not, however, guarantee the stability of the closed-loop system according to the MPC literature [93].

6.5.4   AAQC Attractivity Constraint

The AAQC of the form $\xi_{is}(k, \tau) \geqslant \gamma_{is}\, \xi_{is}(k-1, \tau)$ is expressed as an inequality constraint w.r.t $u_i(k)$ as follows:

$$u_i(k)^T Q_{ic} u_i(k) + 2y_i(k)^T S_{ic} u_i(k) + y_i(k)^T R_{ic} y_i(k) + \sum_{j=k-\tau}^{k-1} \xi_i\left(y_i(j), u_j(i)\right)$$

$$\geqslant \gamma_{ia}\, \xi_{ia}(k-1, \tau). \tag{6.39}$$

This inequality can be rewritten as

$$u_i(k)^T Q_{ic} u_i(k) + 2\Upsilon_{ic}(k) u_i(k) + \delta_{\tau ic}(k) \geqslant 0, \tag{6.40}$$

where $\Upsilon_{ic}(k) := y_i(k)^T S_{ic}$, $\delta_{\tau ic} := y_i(k)^T R_{ic} y_i(k) + \sum_{j=k-\tau}^{k-1} \xi_i\left(y_i(j), u_j(i)\right) - \gamma_{ia}$ $\xi_{ia}(k-1, \tau)$.

The decentralised MPC problem is now re-formulated with the AAQC in the following:

$$\min_{\hat{u}_i, \, \gamma_i(k)} \gamma_i(k) \tag{6.41}$$

$$\text{subject to} \quad \gamma_i(k) > 0, \quad (6.37),$$

$$\begin{bmatrix} -\gamma_i(k) + \delta_i^k + 2\Upsilon_i^k \hat{u}_i & \hat{u}_i^T \\ * & -\Phi_i^{-1} \end{bmatrix} \preccurlyeq 0, \quad \begin{bmatrix} -\delta_{\Upsilon ic}^k - 2\Upsilon_{ic}(k) M_i \hat{u}_i & \hat{u}_i^T M_i^T \\ * & Q_{ic}^{-1} \end{bmatrix} \preccurlyeq 0,$$

where $M_i := [I_{m_i} \, 0_{m_i} \dots 0_{m_i}]_{1 \times h}$.

6.5.5 PAQC Attractivity Constraint

In this section, we also derive a point-wise constraint in $u(k)$ from the PAQC. By rewriting the PAQC inequality as

$$u_i^T Q_{ic} u_i + 2y_i^T S_{ic}^T u_i + y_i^T R_{ic} y_i + \xi_{is}(k-1) \geqslant 0, \tag{6.42}$$

where $Q_{ic} \prec 0$ and $\xi_{is}(k-1)$ is the historical data, y_i is the current-time output vector, it can be developed into a one-time-step quadratic constraint on the initial control vector $u_i(k)$ of the sequence $\hat{u}_i(k, x_i(k), N)$ for the optimisation problem of MPC. The convexity of the inequality constraint will be ensured if $Q_{ic} \prec 0$.

For avoiding any possible conservativeness, we extend the one-time-step constraint to an N_c time-step constraint, called predictive stability constraint (PSC) here. The state predictions in N_c time steps will be used in this PSC, which is a structured constraint on the predictive control sequence $\hat{u}_i$ in N_c future time steps. Nevertheless, to assure the closed-loop system stability from the conditions stated in Theorems 6.5 and 6.6, it is necessary to shrink the predictive horizon N_c to 1, in a gradual manner. With this shrinking strategy, the conservativeness is avoided, yet the closed-loop system stability is guaranteed.

Consider the accumulative supply rate up until the predictive horizon N_c, $\xi_{is}(k + N_c)$, $N_c \geqslant N$. The accumulation $\xi_{is}(k + N_c)$ is split into two components encapsulating the output and input evolutions in the past and in the future, as follows:

$$\xi_{is}(k + N_c) = \xi_{is}(k-1) + \sum_{j=k}^{k+N_c} \xi_i(y_i(j), u_i(j)). \tag{6.43}$$

The predictive PAQC $\xi_{is}(k + N_c) \geqslant 0$ can then be rewritten as

$$\hat{u}_i(k, x_i(k), N)^T \Psi_{ic} \hat{u}_i(k, x_i(k), N) + 2\Upsilon_{ic}^k \hat{u}_i(k, x_i(k), N) + \delta_{ic}^k \geqslant 0, \tag{6.44}$$

where $\Psi_{ic} = \Gamma_{ic}^T C_{iN}^T R_{icN} C_{iN} \Gamma_{ic} + 2\Gamma_{ic}^T C_{iN}^T S_{icN} + Q_{icN}$,

$$\Upsilon_{ic}^k = r_{ic}^{kT} C_{iN}^T (R_{icN} C_{iN} \Gamma_{ic} + S_{icN}),$$

$$\delta_{ic}^k = r_{ic}^{kT} C_{iN}^T R_{icN} C_{iN} r_{ic}^k + \sum_{j=k}^{k+N_c} \xi_i (y_i(j), u_i(j)),$$

in which $C_{iN} = \text{diag}[C_i]_1^{N_c}$, Γ_{ic} and r_{ic}^k are obtained from Γ_i and r_i^k, respectively, with $N \Leftarrow N_c$.

The following theorem states the stabilisability condition for the coupling-delayed network system with the PAQC-based attractivity constraint (6.44).

Denote $\xi_{is}(0, k + N_c) := \sum_{j=k}^{k+N_c} \xi_i (y_i(j), u_i(j))$.

Theorem 6.7 *Let $\xi_{is}(0, k + N_c) > 0$ and $0 < \alpha < 1$.*

1. *Suppose the following LMIs be feasible in P_i, $P_{\tau i}$, Q_i, S_i, R_i, R_{id}, $Q_{\tau i}$, $S_{\tau i}$, $R_{\tau i}$, Q_{ic}, Q_{id}, S_{ic}, R_{ic} :*

    (a) (6.32), (6.16) (for delay-dependent criterion), or
    (b) (6.28), (6.15) (for delay-independent criterion);

2. *With the resultant Q_i, S_i, R_i, Q_{id}, the following LMIs are feasible in $Q_{\tau i}$, $S_{\tau i}$, $R_{\tau i}$, Q_{ic}, S_{ic}, R_{ic} :*

    (a) (6.32), (6.16) (for delay-dependent criterion), or
    (b) (6.28), (6.15) (for delay-independent criterion);

$$r_{ic}^T(k) C_{iN}^T R_{icN} C_{iN} r_{ic}(k) + \xi_{is}(0, k + N_c) \geqslant 0, \tag{6.45a}$$
$$\Gamma_i^T C_{iN}^T R_{icN} C_{iN} \Gamma_i + 2\Gamma_i^T C_{iN}^T S_{icN} + Q_{icN} \prec 0, \tag{6.45b}$$
$$i = 1 \dots h;$$

Then, the controlled system Σ with any h controls $u_i(k)$, $i = 1, 2, \dots, h$, feasible to (6.44) is IpSS stabilised with the PAQC, provided that $\xi_{is}(k) \leqslant \gamma_{is} \times \xi_{is}(k-1)$ whenever $\xi_i (y_i(k), u_i(k)) < 0$.

Proof This is a direct result of Theorems 6.2, 6.5, 6.3 and 6.6, and the convexity and feasibility of the aforementioned predictive stability constraint (6.44):

1. Dissipativity and Stabilisability: (6.16) and (6.32) (item 1.a) are from the dissipativity criteria and stabilisability condition (delay-dependent criteria) in Theorems 6.2 and 6.5. Similarly, (6.15) and (6.28) (item 1.b) are from the dissipativity criteria and stabilisability condition (delay-independent criteria) in Theorems 6.3 and 6.6.

2. Feasibility: The quadratic constraint (6.44) in $\hat{u}_i$ is convex due to $\Psi_{ic} \prec 0$ which is read by (6.45b) in item 2. The stability constraint (6.44) is always feasible due to (6.45a), e.g. $\hat{u}_i = 0$. Then, the optimisation is always feasible by the quadratic objective function and the non-empty convex feasible region (the constraint regions intersect with each other, and do not shrink to a singleton). ∎

In the above theorem, it is required to solve the LMIs for all subsystems at every time step. Since the large-scale matrix $\mathcal{H}$ is inclusive in those LMIs, the problem becomes a centralised one. The convexity of the stability constraint (6.44) is obtained by (6.45b) without having the negative definiteness of coefficient matrices Q_{ic} as usually required in a dissipation condition. The following corollary provides LMI optimisations that can be solved in a decentralised manner, as a practical version of the above theorem. Only delay-independent LMIs are solicited for in this decentralised strategy.

Corollary 6.1 *Let $\xi_{is}(0, k + N_c) > 0$ and $0 < \alpha < 1$.*

For $\mathcal{N}_i := (P_i, P_{\tau i}, Q_i, S_i, R_i, X_i, Q_{\tau i}, S_{\tau i}, R_{\tau i}, Q_{ic}, Z_{ic}, S_{ic}, R_{ic})$, suppose

1. The following LMI optimisations be feasible in

$$\min_{\mathcal{N}_i, X_i > 0, R_{ic} \prec 0} \quad \text{trace}(X_i) \tag{6.46}$$

subject to (6.28), (6.15) *for* $i = 1, 2, \ldots, h$;

2. With $X_i, Q_{\tau i}, S_{\tau i}, R_{\tau i}$, obtained from item 1. above, the following LMI optimisations be feasible in $\mathcal{M}_i := (P_i, P_{\tau i}, Q_i, S_i, R_i, Q_{ic}, S_{ic}, R_{ic})$:

$$\min_{\mathcal{M}_i, R_{ic} \prec 0,} \quad \text{trace}(R_{ic}) \tag{6.47}$$

subject to (6.28), (6.15),

$$r_{ic}^T(k) C_{iN}^T R_{icN} C_{iN} r_{ic}(k) + \xi_{is}(0, k + N_c) \geq 0, \text{ and} \tag{6.48}$$

$$\Gamma_i^T C_{iN}^T R_{icN} C_{iN} \Gamma_i + 2\Gamma_i^T C_{iN}^T S_{icN} + Q_{icN} \prec 0, \tag{6.49}$$

$$i = 1 \ldots h;$$

Then, the controlled system Σ with any h controls $u_i(k)$, $i = 1, 2, \ldots, h$, feasible to (6.44) is IpSS stabilised with the PAQC, provided that $\xi_{is}(k) \leq \gamma_{is} \times \xi_{is}(k-1)$ whenever $\xi_i(y_i(k), u_i(k)) < 0$.

Proof This is a direct result of Theorem 6.7 by using $Q_i, S_i, R_i, Q_{ic}, S_{ic}, R_{ic}$ determined online from the optimisation (6.47) in this theorem, while the other matrices, $Q_{\tau i}, S_{\tau i}, R_{\tau i}$, are determined off-line from the optimisation (6.46). ∎

The feasibility condition of $\xi_{is}(k) > 0$ whenever $\xi_i(k) \leq 0$ is converted into a recursive feasibility constraint, as follows:

$$\begin{bmatrix} -\rho_i(k) - 2y_i^T(k)S_{ic}u_i(k) & u_i^T \\ u_i & Q_{ic}^{-1} \end{bmatrix} \prec 0, \tag{6.50}$$

where $\rho_i(k) := \gamma_{is} \times \xi_{is}(k-1) + y_i^T(k)R_{ic}y_i(k)$.

The DeMPC optimisation with the PAQC is thus as follows:

$$\min_{\hat{u}_i} \quad \gamma_i(k) \tag{6.51}$$

$$\text{subject to} \quad \gamma_i(k) > 0, \text{ and}$$

$$\begin{bmatrix} -\gamma_i(k) + \delta_i^k + 2\Upsilon_i^k\hat{u}_i & \hat{u}_i^T \\ \hat{u}_i & -\Phi_i^{-1} \end{bmatrix} \preccurlyeq 0, \begin{bmatrix} \eta_i & \hat{u}_i^T M_i^T(j) \\ M_i(j)\hat{u}_i & I_{m_i} \end{bmatrix} \succcurlyeq 0, \; j = 1 \ldots N, \text{ and}$$

$$\begin{bmatrix} -\delta_{ic}^k - 2\Upsilon_{ic}^k\hat{u}_i & \hat{u}_i^T \\ \hat{u}_i & \Psi_{ic}^{-1} \end{bmatrix} \preccurlyeq 0, \begin{bmatrix} -\rho_i^k - 2y_i^{kT}S_{ic}M_i(1)\hat{u}_i & \hat{u}_i^T M_i^T(1) \\ M_i(1)\hat{u}_i & Q_{ic}^{-1} \end{bmatrix} \prec 0.$$

The online updating activities for the PAQC and the DeMPC optimisation are summarised in the following procedure:

Procedure 6.1 The off-line computed parameters Γ_i, Θ_i, Φ_i are determined from the state-space realisation matrices A_i, B_i, the weighting matrices $\mathscr{Q}_{iN}^x$ and $\mathscr{Q}_{iN}^u$ of the objective function. Select $\gamma_{is} < 1$ and initiate $N_c \geqslant N$.

1. At step $k = 1$:

   a. Parameter computation:
      i. Initiate $\xi_{is}(0) \gg 0$, $x_i(1)$, $\varepsilon_i(1)$;
      ii. Determine Υ_i^1 and δ_i^1 from Θ_i, Γ_i, the weighting matrices $\mathscr{Q}_{iN}^x$, $\mathscr{Q}_{iN}^u$ of the objective function and $x_i(1)$;
      iii. Solve
         • (6.32) and (6.16) (delay-dependent criterion), or
         • (6.28) and (6.15) (delay-independent criterion)
         for the dissipativity matrices P_i, $P_{\tau i}$, Q_i, S_i, R_i, R_{id}, $Q_{\tau i}$, $S_{\tau i}$, $R_{\tau i}$. Then solve them in combination with (6.45a) using $x_i(1)$ and $\xi_{is}(0)$ for the multiplier matrices Q_{ic}, S_{ic}, R_{ic}. Alternatively, the LMI optimisation (6.46) is solved for delay-independent criteria.
         iv Calculate Υ_{ic}^1, δ_{ic}^1 and ρ_i^1 from Θ_i, Γ_i, the dissipativity matrices Q_{ic}, S_{ic}, R_{ic}, the initial $\xi_{is}(0)$ and $x_i(1)$ as per (6.44) and (6.50);
   b. Formulate and solve the optimisation (6.51) with the calculated parameters Φ_i, Ψ_{ic}, $\tilde{\Psi}_{ic}$, Δ_{ic}, Υ_i^1, δ_i^1, Υ_{ic}^1, δ_{ic}^1 and ρ_i^1 to yield the local control sequence $\hat{u}_i(1, x_i^1, N)$. The first vector element $u_i(1)$ is then applied to control the corresponding subsystem $\mathscr{S}_i$.

2. At step $k \geqslant 2$,

   a. Verify $\xi_i(y(k), u_i(k)) > 0$. If true, omit the stability constraint. Solve the MPC optimisation. Output the first vector element $u_i(k)$ to the corresponding subsystem $\mathscr{S}_i$ and re-iterate this step. Otherwise, go to step (b).

b. Update online parameters with $x_i(k)$:
   i Calculate $\xi_{is}(k-1)$ and update Υ_i^k and δ_i^k;
   ii Solve
   - (6.32) and (6.16) (delay-dependent criterion), or
   - (6.28) and (6.15) (delay-independent criterion)

   in combination with (6.45a) using $x_i(k)$ and $\xi_{is}(0, k + N_c)$ for the multiplier matrices Q_{ic}, S_{ic}, R_{ic}, using the resultant dissipativity matrices in (1.a.iii). Alternatively, the LMI optimisation (6.47) is solved in a decentralised manner for delay-independent criteria.
   iii Update Υ_{ic}^k, δ_{ic}^k and ρ_i^k using Θ_i, Γ_i, the dissipativity matrices Q_{ic}, S_{ic}, R_{ic}, $\xi_{is}(k-1)$ and $x_i(k)$ as per (6.44) and (6.50);
   c. Then, formulate and solve the optimisation (6.51) with its parameters Φ_i, Ψ_{ic}, $\tilde{\Psi}_{ic}$, Δ_{ic}, Υ_i^k, δ_i^k, Υ_{ic}^k, δ_{ic}^k and ρ_i^k to yield the local control sequence $\hat{u}_i(k, x_i^k, N)$. Output the first vector element $u_i(k)$ of $\hat{u}_i(k, x_i^k, N)$ to the corresponding subsystem $\mathscr{S}_i$.
   d. If $N_c > 1$, reduce N_c by one time step, i.e. $N_c - 1$ is used instead of N_c.
   e. Return to Step 2(a).

A numerical example for the AAQC- and PAQC-based attractivity constraints for the DeMPC is given in the next section.

6.6 Illustrative Examples

6.6.1 DeMPC with AAQC

In this section, simulation results for a network system consisting of three coupled subsystems that suffer from a bounded coupling delay are presented to demonstrate the effectiveness of the proposed decentralised MPC strategy with the AAQC-based attractivity constraint. Consider the following continuous-time state-space models of three subsystems with the coupling delay occurring within seven time steps, i.e. $\tau_{max} = 7$ steps. The output constraints are ignored to emulate the system instability. The system model is described by the following matrices:

$$A_1 = \begin{bmatrix} -1 & 0.6 \\ 1 & -0.7 \end{bmatrix}, \quad B_1 = \begin{bmatrix} 0.8 \\ 0.1 \end{bmatrix}, \quad E_1 = \begin{bmatrix} 0.1 \\ 0 \end{bmatrix}, \quad C_1 = \begin{bmatrix} 0.2 & -0.7 \\ 1.2 & 3.1 \end{bmatrix},$$

$$A_2 = \begin{bmatrix} -0.6 & 0 \\ 1 & -0.4 \end{bmatrix}, \quad B_2 = \begin{bmatrix} 1.2 \\ 0.2 \end{bmatrix}, \quad E_2 = \begin{bmatrix} 0.1 \\ 0 \end{bmatrix}, \quad C_2 = \begin{bmatrix} 1.9 & 0.1 \\ -0.1 & -1.5 \end{bmatrix},$$

$$A_3 = \begin{bmatrix} -0.8 & 0 \\ 0.4 & -0.3 \end{bmatrix}, \quad B_3 = \begin{bmatrix} 1.5 \\ 0.1 \end{bmatrix}, \quad E_3 = \begin{bmatrix} 0.1 \\ 0 \end{bmatrix}, \quad C_3 = \begin{bmatrix} -1.5 & -0.1 \\ 0.5 & 1.3 \end{bmatrix},$$

$$F_1 = \begin{bmatrix} 0.1 & 0.2 \end{bmatrix}, \quad F_2 = \begin{bmatrix} 1 & 2 \end{bmatrix}, \quad F_3 = \begin{bmatrix} 0.5 & 4 \end{bmatrix}, \quad H = \begin{bmatrix} 0 & 0 & 1 \\ 1 & 0 & 0 \\ 0 & 1 & 0 \end{bmatrix}.$$

The control and state weighting coefficients and constraints on the manipulation variables are as follows:

$$\mathcal{Q}_1^u = 2, \quad \mathcal{Q}_1^x = \begin{bmatrix} 1 & 0 \\ 0 & 1 \end{bmatrix}, \quad \eta_1 = 5,$$

$$\mathcal{Q}_3^u = 0.5, \quad \mathcal{Q}_3^x = \begin{bmatrix} 0.1 & 0 \\ 0 & 0.1 \end{bmatrix}, \quad \eta_3 = 5.$$

$$\mathcal{Q}_2^u = 0.5, \quad \mathcal{Q}_2^x = \begin{bmatrix} 0.1 & 0 \\ 0 & 0.1 \end{bmatrix}, \quad \eta_2 = 5,$$

The initial control and state vectors of three subsystems are as follows:

$$u_1 = \begin{bmatrix} -0.1 \end{bmatrix}, x_1 = \begin{bmatrix} -3 \\ -4 \end{bmatrix}, u_2 = \begin{bmatrix} -0.5 \end{bmatrix}, x_2 = \begin{bmatrix} 0.2 \\ 0.3 \end{bmatrix}, u_3 = \begin{bmatrix} -1 \end{bmatrix}, x_3 = \begin{bmatrix} 0.4 \\ 4 \end{bmatrix}.$$

The simulation starts with the formulations of three model predictive controllers in SDP without any stability constraints. The predictive horizon of 16 time steps is implemented for all three subsystems, using the Matlab LMI toolbox. The output time responses are shown in Fig. 6.2, which clearly show instability.

To demonstrate the efficacy of the AAQC stability constraint, only off-line computed multiplier matrices are used in this simulation study. The results are illustrated as follows: When the predictive horizon of 12 time steps is implemented for all subsystems, the output time responses are presented in Fig. 6.3a. When the predictive horizon of 16 time steps, the output time responses without the input disturbance are depicted by Fig. 6.3b. This figure shows that the control performance with a longer predictive horizon is better than that with a shorter predictive horizon in Fig. 6.3a. The output time responses for the 16 step horizon case having a bounded input disturbance are shown in Fig. 6.3c, also showing the stabilisation around a neighbourhood of the zero equilibrium.

The average optimising cost trend with input disturbances is provided in Fig. 6.4. The implementation has shown that the chosen predictive horizon plays an important role in the decentralised MPC strategy implemented with the AAQC-based attractivity constraint.

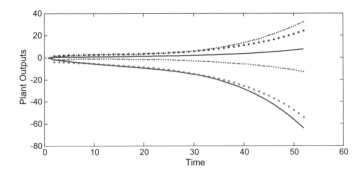

Fig. 6.2 Decentralised MPCs without stability constraints

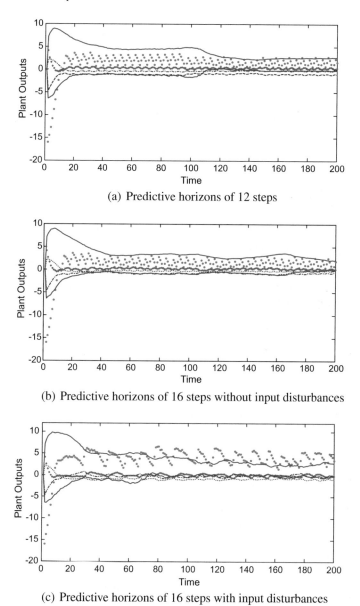

(a) Predictive horizons of 12 steps

(b) Predictive horizons of 16 steps without input disturbances

(c) Predictive horizons of 16 steps with input disturbances

Fig. 6.3 Decentralised MPCs with AAQC using delay-dependent criteria, local input disturbance $\|d_i(k)\| \leqslant 1$

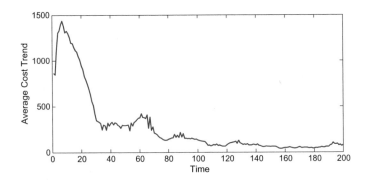

Fig. 6.4 The average optimising cost trend. The coupling delay time $\tau_{max} = 7$ steps

6.6.2 DeMPC with PAQC

The MPC weighting coefficients and constraints on the manipulation variables remain unchanged, as follows:

$$\mathscr{Q}_1^u = 2, \quad \mathscr{Q}_1^x = \begin{bmatrix} 1 & 0 \\ 0 & 1 \end{bmatrix}, \quad \eta_1 = 5,$$
$$\mathscr{Q}_2^u = 0.5, \quad \mathscr{Q}_2^x = \begin{bmatrix} 0.1 & 0 \\ 0 & 0.1 \end{bmatrix}, \quad \eta_2 = 5, \quad \mathscr{Q}_3^u = 0.5, \quad \mathscr{Q}_3^x = \begin{bmatrix} 0.1 & 0 \\ 0 & 0.1 \end{bmatrix}, \quad \eta_3 = 5.$$

The initial state vectors of three subsystems are also the same, i.e.

$$x_1 = \begin{bmatrix} -3 \\ -4 \end{bmatrix}, \quad x_2 = \begin{bmatrix} 0.2 \\ 0.3 \end{bmatrix}, \quad x_3 = \begin{bmatrix} 0.4 \\ 4 \end{bmatrix}.$$

The trajectories of DeMPC with PAQC (Procedure 6.1) corresponding to those from Figs. 6.3 and 6.4 are provided in Figs. 6.5. The results are interpreted as follows: When the predictive horizon of 12 time steps is implemented for all subsystems, the output time responses are depicted in Fig. 6.5a. When the predictive horizon of 16 time steps, the output time responses without the input disturbance are depicted by Fig. 6.5b, showing a better convergence than Fig. 6.3a. The time response with bounded disturbances is given in Fig. 6.5c. The system has been stabilised around a neighbourhood of the zero equilibrium. The average optimising cost trend with input disturbances is also provided, as in Fig. 6.5d.

The implementation has shown that the decentralised MPC strategy that employed the predictive state vectors in the PAQC-based attractivity constraint delivers a better control performance than those from the AAQC-based attractivity constraint.

Fig. 6.5 Decentralised
MPCs with PAQC using
delay-dependent criteria,
local input disturbance
$\|d_i(k)\| \leqslant 1$

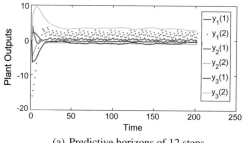

(a) Predictive horizons of 12 steps

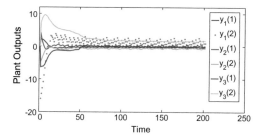

(b) Predictive horizons of 16 steps without input disturbances

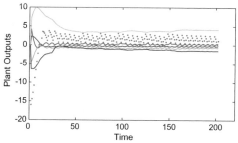

(c) Predictive horizons of 16 steps with input disturbances

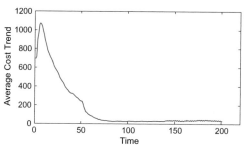

(d) The average optimising cost trend.

6.7   Concluding Remarks

In this chapter, we have presented a decentralised model predictive control strategy for interconnected systems having a coupling delay element. The asymptotically accumulative quadratic constraint (AAQC) has been developed into an attractivity constraint for the decentralised model predictive control of delay coupled interconnected systems. The predictive states are not employed in the AAQC-based attractivity constraint for the DeMPC.

As an alternative, a predictive stability constraint (PSC) has been obtained from the positively accumulative quadratic constraint (PAQC). By employing the state prediction, the conservativeness of a one-time-step constraint that may affect the overall control performance of systems having coupling delays is avoidable while preserving the recursive feasibility.

The accumulative dissipation-based constraints with state predictions in the chapter have been determined as well suited to the interconnection nature of network systems having coupling delays.

Chapter 7
Dependable Control Systems with Internet of Things

This chapter presents an Internet of Things (IoT)-enabled dependable control system (DepCS) for process control systems and manufacturing plants. The hardware system and architecture are also addressed herein, but not only the control design and synthesis as in previous chapters. In a DepCS, the actuator and transmitter form a regulatory control loop without having a separate processor acting as a controller. The processor inside each actuator and transmitter is designed as a computational platform implementing the feedback control algorithm. The connections between the actuators and transmitters via the IoT connectivity create a reliable backbone for the duty and standby controllers of a DepCS, wherein only one controller is responsible for manipulating the plant at any one time. Figure 7.1 shows typically a plant with a DepCS having four controllers.

The centralised input and output marshalling subsystem, as usually required by the other industrial computerised control systems, is not required in a DepCS. A state-feedback control design method for the DepCS applying the self-recovery quadratic constraint is presented in the second part of the chapter. For illustration, an isolated wind–diesel power system is simulated in MATLAB to demonstrate the control design with a DepCS, and a solar tracking system is implemented to show the hardware practicality of an IoT-based DepCS.

7.1 Introduction

Manufacturing and processing plants are predominantly operated by the specialised computers nowadays. These specialised computers and their variants make up the market of industrial computerised control systems. Distributed control systems (DCSs) together with programmable logic controllers (PLCs) are mainly accounted for the market. The control algorithms in discrete logic with Boolean variables are

© Springer Nature Singapore Pte Ltd. 2018
A. Tri Tran C. and Q. Ha, *A Quadratic Constraint Approach to Model
Predictive Control of Interconnected Systems*, Studies in Systems,
Decision and Control 148, https://doi.org/10.1007/978-981-10-8409-6_7

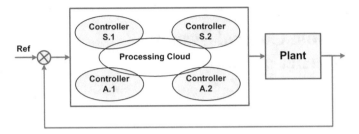

Fig. 7.1 A dependable control system (DepCS) with four controllers

usually installed in a PLC system, while those for the discrete-time regulatory control loops with continuous variables are often installed in a DCS. Both DCS and
PLC systems are emerging as multiple purpose platforms lately for both discrete
control and discrete-time feedback control thanks to the improved performance of
these modern industrial computers. This chapter targets the industrial computerised
control systems for regulatory control loops in the processing plants that may have
hundreds or even thousands of such loops. For conciseness, the term feedback control
system, or simply control system, is used to address a regulatory control loop in this
chapter, but not that in the DCS term. A DCS is often a distributed computer system
that is implemented with different control algorithms, hence not the control system
as we use here.

7.1.1   Industrial Computerised Control System

DCS, which can also be named as process control system (PCS), is a high-integrity
and fault-tolerant distributed computer system with fast real-time control performance and standardised peripheral interfaces. The dependability specification for a
DCS is quantitative and usually much higher than that of an office computer system.
The dependability is achieved by redundant components with duty-standby architecture and online switching-over capability. This is a universal approach for satisfying
the dependability requirement of a fault-tolerant computer system [136].

Dependability here implies the reliability and availability of a system in operation. The quantitative reliability of a system is measured by its probability of being
available and functioning without errors. The quantitative reliability of a system is
specified by its *integrity level* (IL). The integrity level of 99.9% indicates that the system could possibly (and probably) be malfunctioning or unavailable due to failures
in 8.76 hours per year ($0.001 \times 365 \times 24 = 8.76$) while operating. The quantitative
reliability is crucial for many industries, and IL is an important metric in designing
dependable systems. The integrity level of a component is usually calculated from
the mean time between failure (MTBF) and the mean time to repair (MTTR) data
using formulas from industry standards such as IEC 61508 [46]. It is noting that IL

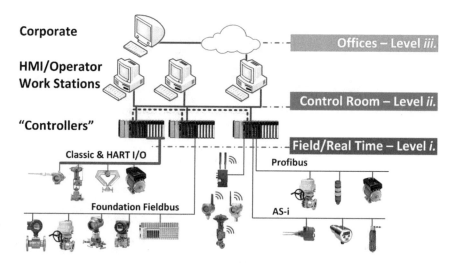

Fig. 7.2 A typical three level DCS structure

is fundamentally different to the safety integrity level (SIL) defined for functional safety systems.

From the engineering perspective, a typical DCS layout can be drawn as in Fig. 7.2 with three tiers of networking, starting from field transmitters and devices, to operator workstations in the central control room for plant operation and management, and up to business planning and corporate layer. This type of DCS layout can be found in well-established standards, such as those from the American Petroleum Institute (API) or the International Society of Automation (ISA). A DCS is designed to accommodate several feedback controllers in its fault-tolerant distributed computer system. The DCS processors, labelled as 'Controllers' in Fig. 7.2, and their input/output (I/O) cards or modules are usually installed in a centralised equipment room next to the control room where the operators interface with and run the plant via computer screens.

The graphic system and their database which link with these 'Controllers' are usually called human–machine interface (HMI) system. The HMI system is also often installed inside the central control room. A block diagram showing field junction boxes, marshalling cabinets, I/O modules and processors is provided in Fig. 7.3. The I/O and marshalling subsystems as well as the equipment room are not required under the proposed control architecture, as explained in the next section.

7.1.2 Internet of Things as Another Step in the Advances of Process Automation

An envisage path for the application of Internet of Things (IoT) in the process automation from our own perspective is illustrated by Fig. 7.4. The circles with the character

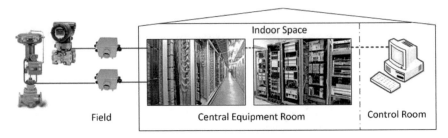

Fig. 7.3 Loop connections in a DCS with junction boxes, marshalling and system cabinets, as well as central control room

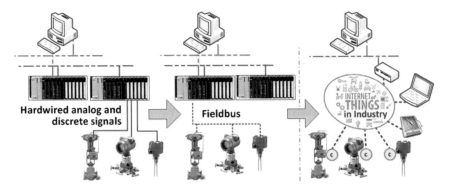

Fig. 7.4 Industrial IoT as another step in the advances of process automation

'c' next to the smart sensor, transmitter and actuator in this figure represent the additional computational capability of the IoT-enabled smart devices, on top of their existing 'smart' functionalities currently available.

With this vision, the needs for a centralised processing capability of the traditional DCSs, as displayed in Fig. 7.2, will vanish in the new system. The loop connections represented in Fig. 7.3 are now simplified without using the I/O and marshalling subsystems as shown in Fig. 7.5. The term *IoT in industry* is used to distinguish with the IoT for public accesses in the corporate environments. It is widely perceived as Industrial Internet of Things (IIoT). The IIoT-enabled connectivity also displaces the 'fieldbus' subsystems, as illustrated in Fig. 7.4. From the device networking perspective, IoT can be considered as another step of fieldbuses. The 'fieldbus' has been implemented with different proprietary protocols developed by different vendors. Therefore, it has been evaluated as fragmented and is limited in data capacity and communication speed. The hardware infrastructure cost for such proprietary fieldbus is also relatively high.

From the information flow and control room perspective, one can also view it as a 'DCS telemetry architecture', since the presenting work is centred on the first layer

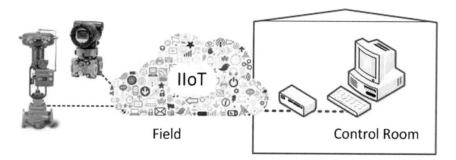

Fig. 7.5 Loop connections in a DSC system enabled by IIoT

for feedback control in the field. The role of the control room from the HMI and operational point of views remains unchanged except for some remote accesses to the field instrumentations and devices for configuring the regulatory control loops which are, now, in the field. The communication protocols of the IoT infrastructure will manage the data flows between the control room and the fields. As we focus on the feedback control synthesis method, the engineering details of IoT systems are not under the scope of this chapter. The IoT is among the chosen technologies in the ongoing 'Smart City' projects around the globe. An organisation called IIoT Consortium has been founded to promote and test rival technologies for industrial applications.

7.1.3 Industrial Standards

The industrial standards for the application software development and configuration of a computerised control system are well applicable to the presenting system since the overall hardware system architecture is simplified with the IoT infrastructure. And the hierarchical and heterogeneous software architectures for different timescales and criticality levels of the applications are still in need. The relevant standards and guidelines include, but are not limited to, ISA 88 for batch processes [168], ISA 95 and IEC 62264 for developing the interfaces between enterprise and control systems and their integration—the manufacturing execution systems [44], IEC 61499 for distributed software architecture [178] and the guidelines for security implementations in the industrial systems such as NIST 800-82.

7.1.4 Research in IoT Security

With the IIoT, automation processes should be less vulnerable to unexpected accesses and malicious attacks, as well as able to guarantee the quality of services and other communication performances for real-time applications. The measures for IoT security have been developed in the computer science and information technology (IT) field; see, e.g. [45, 105] and references therein. Some recent surveys on the protocols, applications and market of IoT can be found in [49, 97, 109]. The IoT security is still in its infancy and is a current research topic in the computer science field, as well as in the industrial research labs. The landscape of this field is expected to grow tremendously in the next five to ten years. We will, therefore, leave it outside the scope of this chapter. An important message from this field is that the users should employ a holistic approach to IoT security, but not only focus on hardware and/or software solutions, while the designers should provide products and solutions that create an underlying end-to-end trust system for the IoT enabled platforms.

The contributions of this chapter can be summarised as follows: firstly, an IoT-based architecture for the DCS is presented to avoid the unnecessary system hierarchies from the legacy designs. Not only the operational functionalities are emphasised, but the reliability and availability of the new architecture are also addressed. Secondly, the DepCS controllers can work in harmony with data exchanged via the IoT and eliminate the inefficient summation of the outputs in the classical approach, designed for analogue and hard-wired signals processed by a limited capacity of the computational platform in the past. Thirdly, with the fully autonomous DepCS, the plug-and-play design for the future control systems becomes feasible. By the same token, the costly input/output and marshalling systems are no more required, which will make the future control system much more cost-effective to manufacture, install, implement and operate. Finally, the control synthesis employing the self-recovery constraint for incrementally dissipative systems is effective in real time, as it is an event-based triggered approach that only re-computes the feedback gain during the duty-standby transitions while static gains are merely used in the remaining time intervals.

This chapter is organised as follows. The DepCS structure and the new DSC system architecture are presented followed by the state feedback control method applying the dissipativity constraint for DepCS synthesis. Attractivity conditions and computation procedure for the state feedback gain are provided. Numerical simulation with small-signal models of the hybrid wind–diesel power system is provided to illustrate the DepCS design. Hardware and experimental results of a DepCS for a solar tracking system are given to exemplify its implementation.

7.2 Dependable Control Systems

The current approach for ensuring the continuous operation of feedback control systems implemented on duty-standby computer systems is to employ the technology of reliable control systems from the control literature; see, e.g. [8] and references therein. A block diagram of such reliable control systems is depicted by Fig. 7.6. This structure is not only bulky but also becomes difficult to accommodate the clustered

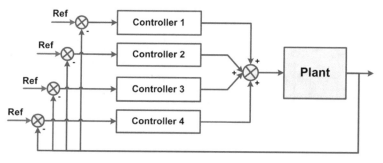

Fig. 7.6 A classical reliable control system (RCS)

topology, peer-to-peer communication and cloud-based applications, which have been evaluated as adequate for use with IoT.

The alternative architecture for the reliable control system of Fig. 7.6 is a dependable control system presented in Fig. 7.1 at the beginning of this chapter. In a processing plant, there may be several regulatory control loops and thus several DepCSs. They form a dependable self-recovery control (DSC) system, as has been designated in [152, 155]. Four processors are assumed to involve in this type of DSC, attributed to the achievable integrity level and available products.

According to industrial data in the computerised control system field, it is usually expected that the regulatory control loops should achieve IL-2 of the range 99.99–99.999% as a minimum in their design specifications. IL-2 will eventually lead to a dual-redundant architecture, as a minimum, if COTS (commercial-off-the-shelf) components are used in the design [46, 136]. The architecture of four duty-standby controllers communicated via the sensor and actuator network is the skeleton of this work to achieve IL 2, or higher, with wired-line or wireless COTS components. By targeting general-purpose components in the architecture, the result will not be limited within a proprietary application but outreaches all standardised products currently available.

7.2.1 Operational Description

Each processor in a DepCS will run the control algorithm independently. They are denoted as controllers S.1/2 and A.1/2 as labelled in Fig. 7.1. Only one of these controllers is active, as a duty controller, at any one time. Therefore, the location of the duty processor varies from time to time. The active program that manipulates the control variables will relocate among these four controllers. We thus call it a processing 'cloud' or 'fog', as it is location independent and has a small number of platforms

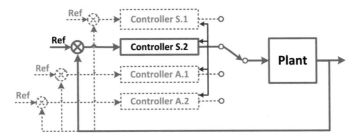

Fig. 7.7 Active and inactive connections in DepCS

(which has a different meaning to the enterprise cloud). The processors can be integrated into the currently used smart transmitters/sensors or actuators. In general, the number of installed processors will depend on the requirements of dependability, operability and cost-effectiveness from a particular application. While the status information is consecutively exchanged between the processors, among those, one acts as a duty processor, the others are in standby, the controller inside each processor will operate in an anonymous manner. On top of the status information, which is managed by the underlying operating system, only one scalar variable is required to be exchanged between the redundantly backed-up controllers. A standby controller can be activated into the duty role relying on this received scalar variable from the duty controller and the trigger signal from the operating system, as represented by the active links between the controllers in Fig. 7.7.

7.2.2   Selection of Controller in Duty Mode

The selection of a controller in the duty mode, which is scheduled by the processor operating system, will be counted upon the healthy state of its hardware, the communication between processors, diagnostic information or user decision. With the clustered oriented structure shown in Fig. 7.1 and a given number of participants in the 'cloud', it can also be called a 'fog'- based system.

7.2.3   Backup Management for Fault-Tolerant Operation

The key for a successful implementation of dependable systems rests with the amount of data to be transferred between the duty and standby components [136]. Fewer amounts of exchanged data will demand less inter-communication over the processor 'cloud' or 'fog'. The presented method requires only one real number to be exchanged between the peer controllers, while the reliable control system of Fig. 7.6 needs four real numbers. And for multi-variable systems, it is even worse as requires the value

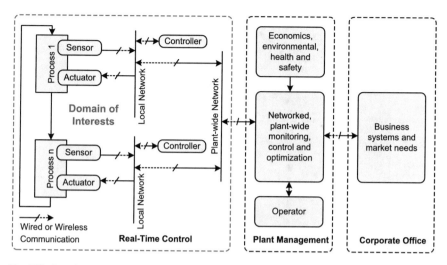

Fig. 7.8 A typical DCS architecture, extracted from [22]

of four vectors to be summed up for the main control vector. More importantly, it is not necessary to exchange this variable at every updating time steps in a DepCS. By virtue of this one-variable approach, which also accepts intermittent data losses, the success of the employed method is assured for both wired-line and wireless networking systems. The IIoT based DSC system is not only energy-efficient, but also achieves a higher dependability owing to the simplicity of exchanging only one variable in both single- and several-variable constrained systems. With these advantages, the new DSC system will be able to have the controllers implemented in a fully decentralised architecture. From this reasoning, we have come up with a new system structure to be discussed in the following subsections.

7.2.4 IoT-Enabled Dependable Self-recovery Control System

7.2.4.1 Existing Distributed Control System

A typical DCS architecture in industry extracted from [22] is shown in Fig. 7.8, wherein the real-time control layer consists of several regulatory control loops. The dependability of the lowest level in this DCS architecture consisting of sensors, actuators and controllers is paramount by virtue of real-time performances. The currently used DCS is usually a fault-tolerant computer network and system. The main processors are implemented with several controllers connecting with sensors and actuators and physically installed in a central control room, or in a few satellite control rooms interconnected with proprietary networks.

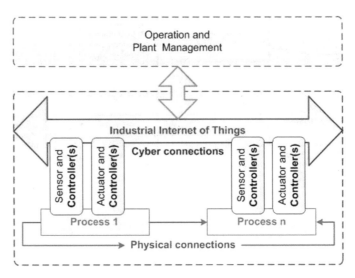

Fig. 7.9 A DepCS and IoT-based architecture. The controllers are integrated into smart sensors and actuators

7.2.4.2 Dependable Self-recovery Control System

The centralised control architecture will vanish when a dependable self-recovery control (DSC) system is installed. The new DSC system will support the plug-and-play (PnP) functionalities, which will lower the installation, testing and operational costs. It will facilitate in the controller installation in the field and on site, far away from the central control room, and in a fully decentralised architecture of IIoT-based networks.

With a DSC system, which uses IIoT-enhanced actuators and transmitters and a novel DepCS structure, the centralised processing and input–output (I/O) marshalling subsystems are not required. The marshalling and I/O system cabinets will thus be eliminated in this new IIoT-enabled DSC system.

The architecture of a fault-tolerant DSC system is shown in Figs. 7.9 and 7.10. The new architecture is much simpler than that of the traditional DCS, whereas the 'Controllers' shown in Fig. 7.2 are not installed. In Fig. 7.9, the controllers are integrated into the smart sensors and actuators in a DSC system. And these smart devices are interconnected via the IIoT.

The relevant control literature for this type of IIoT-based DSC systems is in the networked control system (NCS) strand. Research in NCSs has been intensive during the past decade and is quite mature in its own right, see, e.g. [50]. However, the currently developed control methods for NCSs have not addressed thoroughly the

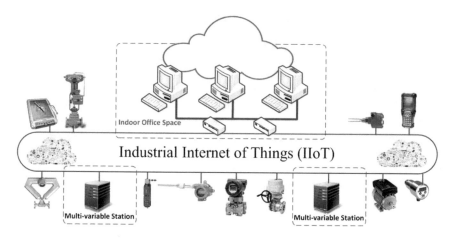

Fig. 7.10 An IoT-enabling distributed computerised control system

dependability mandate for high-integrity applications. The studies in [152, 155] and this chapter propose and partially give a solution to this problem. A control algorithm for DepCSs of the new DSC system is presented in the next section.

7.3 State Feedback Synthesis for Dependable Control Systems

A constrained-state feedback control design method for duty-standby controllers of a dependable control system is presented in this section. As an alternative to the control summation in reliable control systems, only one controller is active at any one time in a dependable control system. The automated managing of duty-standby controllers is challenging, especially in wireless transmitter and actuator networks, owing to the scarcity of both information and processing resources. The solution in this section is effective and feasible, as taking into account both state- and control-incremental constraints, and simply involving a static state feedback with pre-computed and re-computed strategy. The state feedback gains are synthesised to fulfil the strict requirement on the two incremental constraints and simultaneously maintain the control performance such as settling time. As a result of that, the duty-standby controllers will be able to operate independently, while assuring the closed-loop system stability with a newly introduced self-recovery constraint. For a dependable control system, the employed self-recovery constraint is a quadratic constraint with respect to the control and state increments. By satisfying such incremental constraints, the self-recovery constraint will facilitate the independent operation of the controllers while having to exchange only one scalar variable between the redundant controllers. It is an effective method from the information and communication perspective. The

self-recovery constraint packs two information sources, the control and state increments, into one, before transmitting to its peers. The self-recovery constraint at the
receiver side will then unpack the information obtained and use the result for the
local control algorithm.

The incremental passivity approaches have been theoretically presented elsewhere
[106, 129], which imply the usages of signal increments Δu_k and Δx_k in the supply
rate $\xi(\Delta u_k, \Delta x_k)$. The equilibrium-independent passivity has also been introduced in
[60]. Recent developments for the theoretical foundation of differential dissipativity
and incremental stability can be found in [36, 37]. Nevertheless, practical control
applications have not been developed from these theoretical studies. The incremental
dissipativity is employed in this work for the control design problem, which also
incorporates the control- and state-incremental constraints.

7.3.1   Control System Model and Quadratic Constraint

Consider a system $\mathscr{S}$ having a discrete-time state-space model of the form:

$$\mathscr{S} : x(k+1) = Ax(k) + Bu(k), \tag{7.1}$$

where $x(k) \in \mathbb{R}^n$ and $u(k) \in \mathbb{R}^m$ are the state and control vectors, respectively. (A, B)
is controllable. The following control and state constraints are considered herein:

$$\mathbb{U} := \{u : \|u\|^2 \leqslant \eta, \ \eta > 0\}, \tag{7.2}$$

$$\mathbb{X} := \{x : \|x\|^2 \leqslant \rho, \ \rho > 0\}. \tag{7.3}$$

The state increment $\Delta x(k) := x(k+1) - x(k)$, and its constraint is considered
for dependable self-recovery control systems. Specifically,

$$\|\Delta x(k)\|^2 \leqslant \Delta\rho, \tag{7.4}$$

for given $\rho > 0$. Similarly, the control-incremental constraint of the form

$$\|\Delta u(k)\|^2 \leqslant \Delta\eta \tag{7.5}$$

is also inclusive in the problem formulation. Firstly, define a quadratic supply rate
for $\mathscr{S}$, as follows:

$$\xi(\Delta u(k), \Delta x(k)) := \begin{bmatrix} \Delta u^T(k) & \Delta x^T(k) \end{bmatrix} \begin{bmatrix} Q & S \\ S^T & R \end{bmatrix} \begin{bmatrix} \Delta u(k) \\ \Delta x(k) \end{bmatrix} \tag{7.6}$$

where Q, R, S are multiplier matrices with symmetric Q and R. For conciseness, $\xi(\Delta u(k), \Delta x(k))$ is denoted as $\xi_{\Delta(k)}$. The self-recovery quadratic constraint is then defined in the following.

Definition 7.1 Let $0 < \gamma < 1$. The incremental controlled motion $(\Delta u_k, \Delta x_k)$ of $\mathscr{S}$ is said to satisfy the self-recovery quadratic constraint, or simply self-recovery constraint, if the following conditions hold:

$$0 \leqslant \xi_{\Delta(k)} \leqslant \gamma \, \xi_{\Delta(k-1)} \quad \forall k \geqslant 0. \tag{7.7}$$

The incremental dissipativity of $\mathscr{S}$, defined next, plays a more important role in the convergence of $\Delta x(k)$ in this development.

Definition 7.2 The controlled system $\mathscr{S}$ (7.1) is said to be incrementally dissipative with respect to the supply rate $\xi_{\Delta k}$, if there exists a real-valued non-negative storage function $V(\Delta x) := \Delta x^T P_i \Delta x$, $P_i \succ 0$, such that for all $\Delta u(k)$ and all $k \in \mathbb{Z}^+$, the following dissipation inequality is satisfied irrespectively of the initial value of the state increment $\Delta x(0)$:

$$V(\Delta x(k+1)) - \tau V(\Delta x(k)) \leqslant \xi_{\Delta(k)}, \ 0 < \tau < 1. \tag{7.8}$$

For systems having control and state constraints, it is necessary to make some assumptions on the invariance of $\mathbb{X}$, in order for the control problem to be feasible.

Definition 7.3 A set $\Omega \subset \mathbb{R}^n$ is called a constrained control invariant set with respect to $\mathbb{U}$ of the discrete-time system Σ, if for each $x(k) \in \Omega$, $\exists u(k) \in \mathbb{U}$, such that $x(k+1) \in \Omega$ for all $k \geqslant 0$.

Assumption 4 $\mathbb{X} \subset \mathbb{R}^n$ is a constrained control invariant set with respect to $\mathbb{U}$ for the discrete-time system Σ (7.1).

7.3.2   Problem Description

We are concerned with

- The design problem of the constrained-state feedback control law of the form $u(k) = Kx(k)$ for $\mathscr{S}$, such that the closed-loop system (7.1) is stable, subject to the satisfactions of all four constraints on u, x, Δu, Δx, (7.2)–(7.5). However, only soft constraints are considered here since the closed-form memoryless and casual control law will not be able to assure the hard constraints.
- Maintaining the asymptotic property of $\xi_{\Delta(k)}$ among the member controllers of a DepCS. The real-time value of $\xi_{\Delta(k)}$ is transferred between the duty and standby controllers, such that (7.7) is fulfilled by all active controllers. This will be done

by the underlying duty-standby role management mechanism.
- Re-computing the state-feedback gain at every transition event, for the newly assigned duty controller, to assure

   - The state convergence of $\Delta x \rightarrow 0$,
   - The incremental constraint satisfaction and
   - The control performance of a DepCS, such as the settling time and closed-loop stability, is achieved in real time.

   The re-computation is not a persistent online task, as not occurring at every time step, but only at the duty-standby switching-over incidences. The current state vector $x(k)$ is known to the local controller.

Among the above three tasks, the second one will be managed by the operating system of the computer platform running the control algorithm. It is, therefore, an assumption herein. This chapter accomplishes the remaining two tasks of pre-computing and re-computing the feedback gains.

   The asymptotic property of $\xi_{\Delta(k)}$ among the member controllers of a DepCS is crucial in this approach, as it ensures that the control performance of a DepCS is maintained throughout the standby-duty switching-over incidences.

7.3.3   Switching-over Activity and Information

If k_s is the time instant, at which the duty controller is faulty, then $\xi_{\Delta(k_s)}$ is the last known value of $\xi(.)$ to all peer controllers. Here, assume that the switching-over activity will take place in δ time steps, $\delta \geqslant 1$. During the transition time, the last known value of the control vector $u(k_s)$ will be applied to manipulate the plant. This can be done by having a local buffer at the smart actuator or simply using a mechanical latch to keep the actuator at the last position. Once the operating system signalled the completion of the transition, at the time instant $k_s + \delta$, the value of $u(k_s)$ will be retrieved to the newly assigned duty controller, by having $u(k_s) = u(k_s + \delta)$.

   The state-feedback gain pre-computation and re-computation (after each duty-standby switching-over incidence) is delineated in the next section.

7.4   Attractivity Conditions and State Feedback

The sufficient attractivity condition is stated in the theorem below as a basis for the gain computations.

Theorem 7.1 *Let $0 < \tau < 1$, $x(0) \in \mathbb{X}$ and $\xi_\Delta(0) > 0$. Consider the system Σ (7.1). Suppose that Assumption 4 holds, and the closed-loop system Σ and $u = Kx$ is incrementally dissipative with the dissipation inequality (7.8) and fulfils the quadratic constraint (7.7). Then, it is locally asymptotically attractive.*

Proof By applying the asymptotic property of $\xi_{\Delta(k)}$ in (7.7) to the dissipation inequality (7.8), we obtain for every $k \geq 0$,

$$V\big(\Delta x(k+1)\big) \leq \tau V\big(\Delta x(k)\big) + \gamma |\xi_{\Delta(k-1)}|, \ 0 < \gamma < 1.$$

Thus, by iteration

$$V\big(\Delta x_k\big) \leq \tau^{k-1} V\big(\Delta x_1\big) + \gamma |\xi_{\Delta(0)}| (1 + \tau + \cdots + \tau^{k-2}),$$
$$= \tau^{k-1} V\big(\Delta x_1\big) + \gamma |\xi_{\Delta(0)}| \frac{1 - \tau^{k-1}}{1 - \tau}.$$

It is to prove herein that for each $\beta > 0$ there is a finite $k(\beta) > 0$ such that

$$V(\Delta x(k)) \leq \beta \ \forall k \geq k(\beta).$$

Indeed, for each $\beta \geq 0$, there exist two time instants $\bar{k}$ and $\tilde{k}$ such that

$$|\xi_{\Delta(k-1)}| \leq \beta \frac{1 - \tau}{\gamma} \ \forall k > \bar{k} \text{ and } \tau^{k - \tilde{k}} V\big(\Delta x(\tilde{k})\big) \leq \frac{\beta}{2} \ \forall k > \tilde{k}$$

due to (7.7) and $0 < \tau < 1$. Since

$$|\xi_{\iota(k-1)}| \leq \beta \frac{1 - \tau}{\gamma} \ \Rightarrow \ \gamma |\xi_{\iota(\bar{k}-1)}| \frac{1 - \tau^{k-\bar{k}}}{1 - \tau} \leq \frac{\beta}{2},$$

there exists $\check{k} = \max(\bar{k}, \tilde{k})$ such that for each $\beta \geq 0$,

$$V\big(\Delta x(k)\big) < \frac{\beta}{2} + \frac{\beta}{2} = \beta \ \forall k \geq \check{k}.$$

With $P \succ 0$, $\|\Delta x(k)\| \to 0$ as $k \to +\infty$. Applying the result in [36], $x \to x_e$ whenever $\Delta x \to 0$, and $x(k) \in \mathbb{X}$ by Assumption 4, we conclude that the closed-loop system Σ is also locally asymptotically attractive. ∎

There are two feedback gain computations in this approach, the pre-computation in the design phase and the re-computation during the transitions from the standby to the duty controller.

7.4.1 Pre-computing the Feedback Gain

The control law has the state feedback form of $u = Kx$ in this section. The following LMIs are derived from (7.7) and (7.8), by substituting the model of the $\mathscr{S}$ and $u = Kx$ into the corresponding inequalities, rearranging them and applying the Schur complement [14]:

$$\begin{bmatrix} \check{P}_1 & A + B K \\ * & \sigma\check{P}_1 + Y_1 \end{bmatrix} \succcurlyeq 0, \ \check{P}_1 \succ 0, \tag{7.9}$$

$$\begin{bmatrix} \check{M}_1 & A + B K \\ * & \gamma M_1 \end{bmatrix} \succcurlyeq 0, \ M_1 \succ 0, \tag{7.10}$$

where $P_1 := (K^T B^T + A^T) P (A + B K)$, $\quad M_1 := (K^T B^T + A^T) M (A + B K)$, $M := Q + SK + K^T S^T + K^T RK$, $M_1 = P_1 Y_1 P_1$, $\check{M}_1 := M_1^{-1}$, $\check{P}_1 := P_1^{-1}$.

A similar derivation is provided in Appendix A.5.

Similarly, the constraints (7.2)–(7.5) are satisfied when the following LMIs are fulfilled:

$$\begin{bmatrix} I & A + B K \\ * & I \end{bmatrix} \succcurlyeq 0, \ \begin{bmatrix} I & A + B K - I \\ * & \frac{\Delta\rho}{\rho} I \end{bmatrix} \succcurlyeq 0, \tag{7.11}$$

$$\begin{bmatrix} I & K \\ * & \frac{\eta}{\rho} I \end{bmatrix} \succcurlyeq 0, \ \begin{bmatrix} I & K \\ * & \frac{\Delta\eta}{\Delta\rho} I \end{bmatrix} \succcurlyeq 0. \tag{7.12}$$

The computation for K is then as follows: firstly, the matrices $\check{P}_1$, Y_1 and K are found from the solution to (7.9), (7.11) and (7.12) by the optimisation of

$$\max_{\check{P}_1, Y_1, K} x_0^T \check{P}_1 x_0 \ \ s.t. \ (7.9), (7.11), (7.12). \tag{7.13}$$

The objective function of $x^T(0) \check{P}_1 x(0)$ is employed herein to guarantee the performance of $\sum_k^\infty x^T P x$ as usually determined in the control literature. Equation (7.10) cannot be included as it is not an LMI. The matrices P_1 and M_1 are then obtained from $\check{P}_1$, Y_1 and K accordingly.

Secondly, assuming $K^T = P_1^{-1} X^T$ and $M_1 = Z_1 P_1$, X and Z_1 are then calculated from the resultant P_1 and Y_1. Subsequently, P_1 is re-computed off-line by solving the equivalent LMIs of (7.9), (7.11), (7.12). The equivalent LMIs, which have been derived from (7.9), (7.11), (7.12) using Schur complement, are provided below.

$$\begin{bmatrix} \check{P}_1 & A + B X \check{P}_1 \\ * & \sigma \check{P}_1 + \check{P}_1 Z_1 \end{bmatrix} \succcurlyeq 0, \quad \begin{bmatrix} \check{P}_1 Z_1^{-1} & A + B X \check{P}_1 \\ * & \gamma \check{P}_1 Z_1 \end{bmatrix} \succcurlyeq 0, \tag{7.14}$$

$$\begin{bmatrix} I & A + B X \check{P}_1 \\ * & I \end{bmatrix} \succcurlyeq 0, \tag{7.15}$$

$$\begin{bmatrix} I & A + B X \check{P}_1 - I \\ * & \frac{\Delta \rho}{\rho} I \end{bmatrix} \succcurlyeq 0, \tag{7.16}$$

$$\begin{bmatrix} I & X \check{P}_1 \\ * & \frac{\eta}{\rho} I \end{bmatrix} \succcurlyeq 0, \quad \begin{bmatrix} I & X \check{P}_1 \\ * & \frac{\Delta \eta}{\Delta \rho} I \end{bmatrix} \succcurlyeq 0. \tag{7.17}$$

Then, the LMI optimisation of

$$\max_{\check{P}_1 \succ 0} x_0^T \check{P}_1 x_0 \quad s.t. \ (7.14), (7.15), (7.16), (7.17), \tag{7.18}$$

will be solved for $\check{P}_1$. The feedback gain $K = X P_1^{-1}$ is obtained accordingly. This feedback gain will be used in the control law until the occurrence of a duty-standby transition.

7.4.2   Re-computing the Transition Feedback Gain

The control needs to keep track with the evolution of the self-recovery constraint to ensure the closed-loop system stability and to simultaneously maintain the control performance of settling time. When the duty-standby transitions occur, the feedback gain K is necessarily re-computed at the new duty controller based on the received value of $\xi_{\Delta(k_s)}$ from the previous duty controller. It is assumed here that the transition is accomplished at the time step $k_s + \delta$. The newly computed feedback gain K is then applied from the time step $k_s + \delta$ onward, until the occurrence of the next duty-standby transition.

Based on the inequality of the self-recovery constraint (7.7), the current state $x(k)$, the last known value of $\xi_{\Delta(k_s)}$ and the retrieved value of $u(k-1) = u_{(k_s)}$, the following LMI is derived for re-computing the feedback gain K:

$$x_{(k)}^T (A + B K - I)^T M (A + B K - I) x_{(k)} - \gamma \xi_{\Delta(k_s)} \preccurlyeq 0, \tag{7.19}$$

which is equivalent to the LMI below.

$$\begin{bmatrix} M_1^{-1} & x_{(k)} \\ * & \gamma \xi_{\Delta(k_s)} \end{bmatrix} \succcurlyeq 0, \ M_1 \succ 0. \tag{7.20}$$

The above inequality ensures that the constraint on the state increment will be satisfied between the two time steps k_s and k. Similarly, the control-incremental constraint (7.5) also needs to be satisfied, by the following LMI:

$$\begin{bmatrix} I & X \check{P}_1 x_{(k)} - u_{(k_s)} \\ * & \Delta\eta \end{bmatrix} \succcurlyeq 0. \tag{7.21}$$

Using the known values of $x(k)$, $\xi_{\Delta(k_s)}$, $u_{(k-1)} = u(k_s)$, and the resultant X and Z_1 from the pre-computation design phase while assuming that $K^T = P_1^{-1} X^T$ and $M_1 = Z_1 P_1$, we re-compute the gain K online using the LMI optimisation in the following:

$$\max_{\check{P}_1} x_{(k)}^T \check{P}_1 x_{(k)} \tag{7.22}$$

$$s.t. \begin{bmatrix} \check{P}_1 Z_1^{-1} & x_{(k)} \\ * & \gamma \xi_{\Delta(k_s)} \end{bmatrix} \succcurlyeq 0,$$

$$(7.14),\ (7.15),\ (7.16),\ (7.17),\ \text{and}\ (7.21).$$

The feedback gain $K = X P_1^{-1}$ is obtained accordingly. This gain K is then applied to the control law until the next occurrence of a duty-standby transition. The feedback gain computation is summarised as follows:

Procedure 1 *State feedback gain computation for DepCS*

1. *Pre-computation:*

   a. *Solve the optimisation (7.13) for $\check{P}_1$, Y_1 and K.*
   b. *Obtain P_1, M_1, Z_1, X.*
   c. *Solve the optimisation (7.13) for $\check{P}_1$.*
   d. *Obtain $K = X \check{P}_1$.*

2. *Re-computation: Assume that $\xi_{\Delta(k)}$ is transferred between the peer controllers at every step k. At a duty-standby transition step $k_s + \delta$ triggered by the operating system, the newly assigned duty controller will:-*

   a. *Solve the optimisation (7.22) for $\check{P}_1$, using X, Z_1 from the pre-computation, and the known $\xi_{\Delta(k_s)}$, $x_{(k)}$ and $u_{(k-1)} = u_{(k_s)}$.*
   b. *Obtain $K = X \check{P}_1$.*

The simulation studies in MATLAB environment have been conducted with the results being presented in the next section.

Fig. 7.11 A typical isolated wind–diesel power system

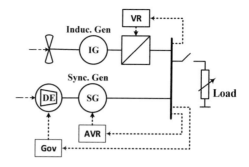

7.5 DepCS Design and Implementation Examples

7.5.1 Isolated Wind–Diesel Power System

The small-signal linearised model of an isolated wind–diesel power system with local PI controllers, taken from [11], has been used in this numerical example. This wind–diesel power system consists of a wind generator and a diesel generator connecting to a common bus bar. The wind generator has a wind turbine, an induction generator and the converter/inverter with its own voltage regulator. The diesel generator set consists of a diesel engine and a synchronous generator with AVR (automatic voltage regulator) and governor, as sketched out by Fig. 7.11. The two state-space realisation matrices A and B of $\mathscr{S}$ (7.1) are provided below. The model parameters are given in [11] with the following state realisation matrices.

$$
A = \begin{bmatrix}
-7.4 & 5 & 0 & 0 & 7.47 & 0 & 0 & 0 \\
0 & -0.333 & 0.333 & 0.333 & 0 & 0 & 0 & 0 \\
-0.02 & 0 & -0.5 & 0 & 0 & 0 & 0 & 0 \\
-1.58 & 0 & 0 & -40.0 & 0 & 0 & 0 & 0 \\
0.374 & 0 & 0 & 0 & -0.623 & 0.25 & 0 & 0 \\
0 & 0 & 0 & 0 & 0 & -1 & .14 & 0.084 \\
0 & 0 & 0 & 0 & 0 & 0 & -1 & 0.4 \\
0 & 0 & 0 & 0 & 0 & 0 & 0 & -24.39
\end{bmatrix}
$$

$$
B = \begin{bmatrix}
0 & 0 \\
0 & 0 \\
\frac{K_D(T_{D2}-T_{D1})}{T_{D2}(T_{D2}-T_{D3})} & 0 \\
\frac{K_D(T_{D3}-T_{D1})}{T_{D3}(T_{D3}-T_{D2})} & 0 \\
0 & 0 \\
0 & 0 \\
0 & 0 \\
0 & \frac{K_{P2}}{T_{P2}}
\end{bmatrix}
=
\begin{bmatrix}
0 & 0 \\
0 & 0 \\
0.101 & 0 \\
7.898 & 0 \\
0 & 0 \\
0 & 0 \\
0 & 0 \\
0 & 24.390
\end{bmatrix}
\quad
L = \begin{bmatrix}
5 & 0 \\
0 & 0 \\
0 & 0 \\
0 & 0 \\
0 & 0.25 \\
0 & 0 \\
0 & 0 \\
0 & 0
\end{bmatrix}
$$

Fig. 7.12 DSC system with self-recovery constraint

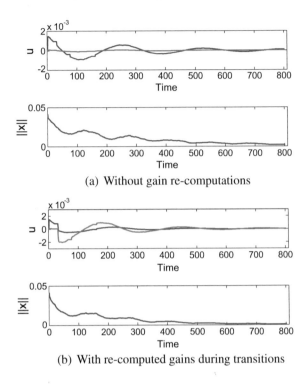

(a) Without gain re-computations

(b) With re-computed gains during transitions

The frequency control problem here is to stabilise the system frequency and diesel generator power in the events of small load changes or wind power variations. The two monitored states are the system frequency and diesel power which are the first and second elements in the state vector, respectively. The updating time is chosen as $\tau_s = 0.1$, while the initial state vector is chosen as $x(0) = 10^{-3} \times [24.5\ 15.7\ -16.3\ -8.7\ 11\ -10\ 11-14]^T$. The constraints are set with $\eta = 2 \times 10^{-4}$, $\rho = 5 \times 10^{-3}$, $\Delta\eta = 2 \times 10^{-5}$, $\Delta\rho = 5 \times 10^{-4}$. The coefficients $\gamma = 0.999$ and $\tau = 0.9999$ have been selected in this simulation study. The pre-computed feedback gain is as follows:

$$K = 10^{-2} \times \begin{bmatrix} 3.20\ 3.42\ 0.55\ 18.80\ 2.58\ 0.65\ 2.31\ 0.48 \\ 0.22\ 0.38\ 0.37\ \ 0.22\ \ 0.43\ 0.41\ 0.41\ 0.61 \end{bmatrix}.$$

In the first simulation study, the re-computed gains are not used in the control algorithm: Only the pre-computed gain is used in the control laws. The control and state trajectories are shown in Fig. 7.12a with three transition events occurring at the time steps 20, 65 and 150. During the transition intervals, the control retains its previous known value. The transitions are assumed taken place in 10 time steps, i.e. $\delta = 10$. The state and control trajectories with the above feedback gain show a stabilised system, but the state does not converge to zero after 800 time steps. A fading disturbance signal which is proportional to the state vector has been added to the model as an input disturbance to show the effectiveness of the self-recovery constraint.

In the second simulation study, the gain is re-computed for the three transition events. The corresponding trajectories are shown in Fig. 7.12b. The control trajectory with the re-computed gains indicates a different performance to that in Fig. 7.12a, using the same disturbance pattern and magnitudes. As a result, the state vector reaches zero after 600 time steps, which is better than that in Fig. 7.12a. The three re-computed gains are provided below for information.

$$10^{-2} \times \begin{bmatrix} 3.04 & 3.06 & 3.40 & 77.53 & 3.16 & 3.17 & 3.10 & 2.11 \\ 2.34 & 2.29 & 14.11 & -0.06 & 1.82 & 13.35 & 6.87 & 28.58 \end{bmatrix},$$

$$10^{-2} \times \begin{bmatrix} 2.82 & 2.91 & 3.05 & 77.59 & 3.01 & 2.85 & 2.82 & 1.94 \\ 2.23 & 1.97 & 14.05 & -0.06 & 1.57 & 13.11 & 6.89 & 28.18 \end{bmatrix},$$

$$10^{-2} \times \begin{bmatrix} 2.68 & 2.78 & 2.85 & 77.63 & 2.91 & 2.64 & 2.67 & 1.82 \\ 2.25 & 1.83 & 14.28 & -0.05 & 1.40 & 13.37 & 7.21 & 27.81 \end{bmatrix}.$$

When there are not any transition events, the state also reaches zero after 600 time steps, which is compatible to those of the DepCS with re-computed feedback gains in Fig. 7.12b. The settling time in the case of applying the re-computed gains is around 600 time steps, which is also the settling time of the case without having duty-standby transition. The re-computation uses the value of $\xi_{\Delta(k_s)}$ transferred from the previous active duty controller. The trend of self-recovery constraints is shown in Fig. 7.13 for the cases with pre-computed and re-computed gains. The latter demonstrates a smoother trajectory and the settling time conservation owing to the gain re-computation using the passing-on value of the self-recovery constraint.

Two other cases of different intervals of transition events are given in Figs. 7.14 and 7.15. Both cases have shown the improved control performances of the DSC with the gain re-computations, evaluated by the accumulative sum of $V = x^T P x$ along the trajectory, as depicted by Figs. 7.14 and 7.15.

The simulation study has demonstrated that the settling time of the DepCS is well maintained in the three events of duty-standby switching-over as a result of applying the self-recovery constraint.

The comparisons between the presented control design and the LQR design are shown in Fig. 7.16, in which both designs have delivered a similar result but the re-computed gains in the DepCS help achieve a better control performance over the LQR. Moreover, the self-recovery constraint design does not require a tuning of the weighting matrices, while the LQR does.

Fig. 7.13 Self-recovery
constraint evolution

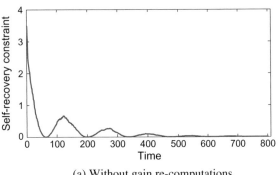

(a) Without gain re-computations

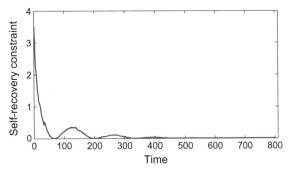

(b) Less and lower peaks with re-computed gains

7.5.2   Solar Tracking Control System for PV Panel

This second example presents the implementation of a DepCS architecture applied
to the problem of solar tracking for a photovoltaic (PV) panel with experiments.
Figure 7.17 shows the overall functional design of the solar energy management
system. The solar tracking module consists of the photovoltaic panel driven by a
motion mechanism with two degrees of freedom for controlling the azimuth and
elevation of the panel, as shown in Fig. 7.18. The motion can be controlled by either
a built-in or a remote controller. A gateway and Modbus/TCP protocols are used
to interface the solar tracking control systems and the PV panel with an energy
management system. A 'Tigo' system is installed to record the solar panel and energy
system status. An application server aggregates the recorded data to analyse the power
usage, evaluate the harvested energy and detect abnormal events. The meteorological
and astronomical data can also be retrieved and integrated into the system to optimise
the energy harvesting process.

A block diagram of the IoT network and the dependable control system is shown
in Fig. 7.19. At the centre of the system are the IoT-enabled dependable controllers,
as depicted by Fig. 7.20. The dependable control system (DepCS) is implemented
to maintain the system reliability and continuity for the operation in the presence of

Fig. 7.14 DSC
system—different transition
time intervals

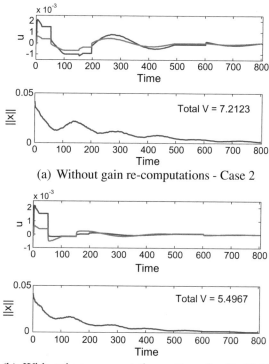

(a) Without gain re-computations - Case 2

(b) With gain re-computations - Case 2 with 23.8%
improvement.

hardware failures and communication interruptions. The control and duty-standby management algorithms are implemented on the embedded computer boards inter-connected over a private cloud network. In this design, an actuator device does not necessarily possess information of which control board it is receiving the control sig-nals from because there are several redundant controllers available with the online switching-over capability. The system thus allows a particular controller to simulta-neously manipulate multiple PV panels.

Different data transport protocols can be chosen in practice: the Transmission Control Protocol (TCP) is a proper choice for the 'slower' plants, while the User Datagram Protocol (UDP) and Real-time Transport Protocol (RTP) may be better options for 'faster' plants. In our experiment, the TCP has been chosen for the solar tracking system. Together with the communication links, the successful implementa-tion of the dependable control system rests with the availability of IoT boards of the embedded computer and the online switching-over capability of the controllers. This allows for one particular controller to be able to control multiple plants while to also

Fig. 7.15 DSC
system—different transition
time intervals

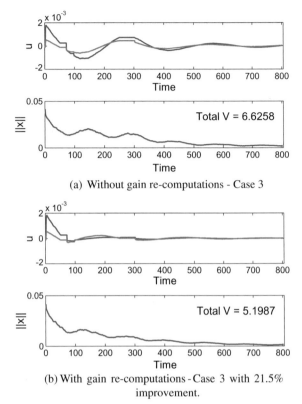

(a) Without gain re-computations - Case 3

(b) With gain re-computations - Case 3 with 21.5%
improvement.

Fig. 7.16 Self-recovery
constraint versus LQR. The
performance with LQR
depends on the choices of
weighting matrices

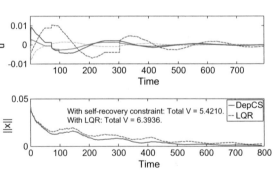

serve as a backup controller, as can be seen in the block diagram of the IoT network
and DepCS of Fig. 7.19. Since only one controller is active as the duty controller,
while the others are in the standby mode, the location of the duty processor varies
from time to time depending on the underlying state machine. We have implemented
a state machine for managing the system operations such that a standby controller
can be activated into the duty mode in the presence of one of the following events:

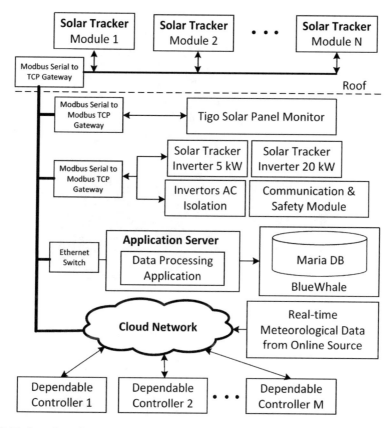

Fig. 7.17 Overview of the solar energy management in a microgrid

1. The duty controller is running out of resources requesting a switch to the standby mode;
2. The duty controller is facing hardware failures releasing a switching-over token;
3. The data communication is interrupted triggering a time-out event;
4. An unknown problem occurs causing zero state variable to be broadcast.

The switching-over process is carried out in a 'first-come, first-served' manner. The first standby controller that catches the released token will be activated into the duty mode and take the permission to broadcast the state variables.

Experiments have been conducted to validate the DepCS design and implementation for the energy management system of a microgrid. Here, the set-up includes

Fig. 7.18 Solar tracker

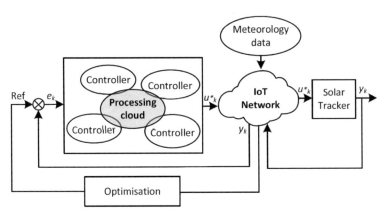

Fig. 7.19 Structure of the implemented dependable control system

a solar tracker and a PV array located at the rooftop of a twelve-storey building of
the University of Technology Sydney, Australia [112]. The solar tracker includes the
hardware interfaces to control its azimuth and elevation within the range of $0°$–$90°$.
The PV array has 72 modules, each has the maximum power of 280 W and the efficiency of 22.7%. The IoT boards consist of a quad-core Intel Atom processor, 2.4
GHz, and a quad-core Broadcom BCM 2837 64-bit CPU, 1.2 GHz, both run Linux
operating system and are connected to the cloud via wireless 802.11 connections. The
controllers are executed in a cascade mode in which the inner loop is a built-in PID

Fig. 7.20 Embedded computers and IoT boards

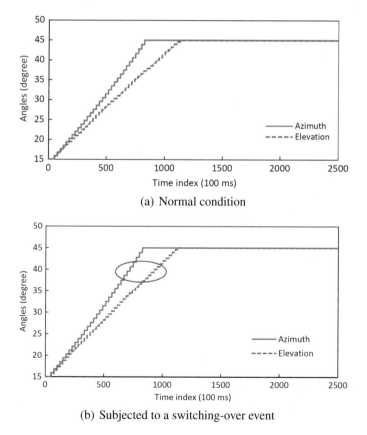

Fig. 7.21 Time response of the solar tracker with the azimuth and elevation set to 45°

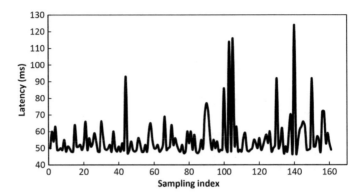

Fig. 7.22 Network latency during the control process

controller, while the outer loop is the DepCS obtained as per the control procedure. The meteorology information for creating the set points is retrieved online with the network service APIs. The inputs are $151.1990°$ E in the longitude and $-33.8840°$ S in the latitude corresponding to the tracker location. The data communication is carried out via TCP protocol with socket programming.

In the first experiment, we have tested the system capability to track the sun direction by providing some angle set points. Figure 7.21a shows the output time response of the sun tracker with $45°$ set point. Both the azimuth and the elevation's angular references are reached without any ripples. The response trajectories are rather linear thanks to the inner PID loop settings. The settling time of 100 s in this experiment meets the requirement of tracking the sun whose direction only changes slightly in several minutes. Figure 7.22 shows the network latency during the experiment time. The average latency of 55 ms is rather small compared to the controller updating time of 100 ms and the settling time of 100 s and thus has not affected the control performance in this experiment.

In the second experiment, the reliability and self-recovery capability of the system were verified when it underwent a hardware failure by intentionally unplugging the power supply of the duty controller during its operation. This action caused interruptions to the exchanged data between the IoT boards and some time-out events. Without having the DepCS with the redundancy management programme and state machine, the system is unable to recover after these incidents. When the DepCS is in place, a standby controller can detect the time-out events and switch over to the duty mode. The 'performance variables' including the set points, state variables and sensor outputs that have been memorised from the previous communication cycles can be used by the in-duty controller. Figure 7.21b shows the output time response of the DepCS. The solar tracker eventually reaches the expected operating angles thanks to the DepCS even when time-out events occur several times. Figure 7.23 shows the

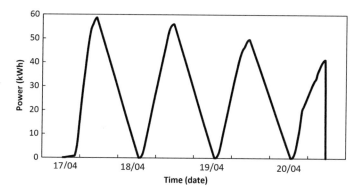

Fig. 7.23 Solar energy generated from 17/04/2017 to 20/04/2017

recorded energy generated by the PV panels in four consecutive days, from 6 am to 5 pm each day. The harvested energy is quite stable during daytime with the average accumulative energy of 48 kWh per day.

7.6 Concluding Remarks

This chapter has presented a dependable self-recovery control (DSC) architecture that congregates the Internet of Things (IoT) connectivity and dependable control systems (DepCSs) as an alternative to the widely perceived hierarchical architecture of the computerised control systems in moving towards new trends in modern industrial automation. The emerging IoT-enabled dependable control systems are capable of accommodating a fully decentralised architecture with wired-line or wireless communication networks, yet meeting the quantitative reliability specifications. The conventional input/output hardware subsystems and the centralised processors are no more required in the IoT-enabled architecture. Also, a state feedback control scheme with the quadratic constraint has been developed for the DepCS in the second part of this chapter.

The state feedback gain is re-computed at every duty-standby switching-over incidence to ensure that the incremental constraint is satisfied and the control performance is maintained. Numerical simulations for an isolated wind–diesel power system has demonstrated the effectiveness of the quadratic constraint for the dependable control systems. In addition, a solar tracking control system with PV panels for a microgrid has been implemented with the DepCS in an experiment with actual IoT boards and wireless communication networks. The experiments have shown reliable operations with successful switching-over activities between the duty and standby controllers of the DepCS.

Appendix A
General Dissipativity Constraint

The asymptotically positive realness constraint (APRC) and quadratic dissipativity constraint (QDC) introduced in Chap. 2 are the special cases of the general dissipativity constraint (GDC) to be presented in this Appendix. The absolute function is employed for the supply rate, and the $\mathcal{KL}$−bounded functions are used for representing the GDC stability. In Chap. 2, the asymptotic attractivity conditions with the quadratic constraints have been stated and applied to the decentralised MPC problem. However, the stability of a controlled system in general has not been fully analysed therein. The stability obtained from the stabilisation with the GDC is included here in the context of Lyapunov stability, Lagrange stability and asymptotic stability. With the advantage of having the storage function behave as a relaxed non-monotonic Lyapunov function, the GDC method is less conservative than the original Lyapunov's method with monotonic Lyapunov functions. The Lyapunov stability is not, nevertheless, assuredly obtained in the GDC method. The GDC provides a type of stability that is similar to the Lyapunov stability starting from a future time instant $k^* > 0$ plus the convergence property. A controlled system with the GDC is said to be *'stable in the GDC sense'* or simply *GDC stable*. The GDC also provides a boundedness property that is similar to the Lagrange uniform boundedness with an extra feasible condition. The *'input-to-power-and-state stability'* (IpSS) is introduced as an extension of the GDC stability for systems having internal and external perturbations, similarly to the input-to-state stability as an extension to Lyapunov stability. Also, a GDC stability condition for constrained controlled systems with the model predictive control is included at the end of this Appendix.

A.1 General Dissipativity Constraint

Consider a discrete-time system $\mathscr{S}$ of the form:

$$\mathscr{S} : x(k+1) = f(x(k)) + B(x(k))u(k) + Ld(k), \qquad (A.1)$$

© Springer Nature Singapore Pte Ltd. 2018
A. Tri Tran C. and Q. Ha, *A Quadratic Constraint Approach to Model Predictive Control of Interconnected Systems*, Studies in Systems, Decision and Control 148, https://doi.org/10.1007/978-981-10-8409-6

where $x \in \mathbb{X} \subset \mathbb{R}^n$, $u \in \mathbb{U} \subset \mathbb{R}^m$ are the state and control vectors, respectively; $f(x)$ is a vector field, $f : \mathbb{R}^n \rightarrow \mathbb{R}^n$; $B(x)$ is a matrix field, $B : \mathbb{R}^n \rightarrow \mathbb{R}^{n \times m}$; the elements of f and B, denoted as $f_{[i]}(x)$; and $B_{[i,j]}(x)$, are not necessarily continuous functions of x, $f_{[i]} : \mathbb{R}^n \rightarrow \mathbb{R}$ and $B_{[i,j]} : \mathbb{R}^n \rightarrow \mathbb{R}$; $d(k)$ represents an unknown disturbance, $d(k) \in \mathbb{R}^q$, but bounded: $\|d(k)\|^2 \leqslant \theta < +\infty$. Without loss of generality, we assume that $\mathbb{X}$ is compact, $0 \in \mathbb{X}$, $0 \in \mathbb{U}$, and $f(0) = 0$.

The general dissipativity constraint (GDC) to be defined in this section is applicable to the discrete-time systems of the form $\mathscr{S} : x(k+1) = f(x(k), u(k)) + Ld(k)$, in general. However, the compound output vector introduced next will be more suitable for the input-affine system (A.1).

The following supply-rate function can be implemented in various engineering problems:

$$\xi_\triangle(x, w) = [x^T \ f(x)^T] Q \begin{bmatrix} x \\ f(x) \end{bmatrix} + 2[x^T \ f(x)^T] Sw + w^T Rw,$$

where Q, S, R are coefficient matrices with appropriate dimensions, $Q = Q^T$ and $R = R^T$. The quadratic form of the supply-rate function is well perceived as the general quadratic supply-rate function for a dissipative system in the control literature, see, e.g. [17, 172]. Such a dissipative system can also be called (Q, S, R)-dissipative system [59]. In some developments, $Q = \alpha I$, $R = \gamma I$, and $S = 0$ may be employed to derive the input-output gains for linear systems.

Next, define the controlled supply rate $\xi(k, x(k), u(k))$, which is also a real-valued piecewise-continuous function in x and u, $\xi : \mathbb{Z} \times \mathbb{R}^n \times \mathbb{R}^m \rightarrow \mathbb{R}$. The initial $\xi_{(0)} := \xi(0, x(0), u(0))$ is finite. For any $x(k) \in \mathbb{X}$, the control sequence $\{u(k) \in \mathbb{U}\}$ is such that the supply rate satisfies the following bounded condition:

$$\exists \theta \in \mathbb{R}^+ : \sum_{k=0}^{\kappa} |\xi_{(k)}| \leqslant \theta \quad \text{for all } \kappa > 0, \tag{A.2}$$

where $\xi_{(k)} := \xi(k, x(k), u(k))$.

Definition A.1 The controlled motion $(x(k), u(k))$ of $\mathscr{S}$ is said to satisfy the general dissipativity constraint (GDC), if there exists a supply rate $\xi(k, x(k), u(k))$ and there is a function $\alpha(.)$ of class $\mathscr{K}\mathscr{L}$, such that

$$\left| \xi(k, x(k), u(k)) \right| \leqslant \alpha(|\xi_{(0)}|, k) \quad \forall k \in \mathbb{Z}^+. \tag{A.3}$$

Definition A.2 The controlled motion $(x(k), u(k))$ of $\mathscr{S}$ is said to satisfy the GDC, practically, if there exists a supply rate $\xi(k, x(k), u(k))$ and there is a function $\alpha(.)$ of class $\mathscr{K}\mathscr{L}$, such that

$$\left| \xi(k, x(k), u(k)) \right| \leqslant \alpha(|\xi_{(0)}|, k) + \varphi(k) \quad \forall k \in \mathbb{Z}^+, \ \varphi(k) > 0. \tag{A.4}$$

Definition A.3 A function $\zeta : \mathbb{R}^p \to \mathbb{R}^p$ is called $\mathcal{K}\mathcal{L}$ bounded, if there exists a class $\mathcal{K}\mathcal{L}$ function $\alpha(.,.)$, such that for all $\zeta(k) \in \mathbb{R}^p$, $k \in \mathbb{R}_0^+$, we have the inequality $\|\zeta(k)\| \leqslant \alpha(\|\zeta(0)\|, k)$.

The above GDC can then be simply stated as follows: The controlled motion $\big(x(k), u(k)\big)$ of $\mathscr{S}$ is said to satisfy the GDC, if there exists a supply rate $\xi\big(k, x(k), u(k)\big)$ satisfying the bounded condition (A.2), that is also $\mathcal{K}\mathcal{L}$-bounded.

Lemma A.1 *Consider $\mathscr{S}$ and the supply rate $\xi\big(k, x(k), u(k)\big)$ – a real-valued piecewise-continuous function of x and u, $\xi : \mathbb{Z} \times \mathbb{R}^n \times \mathbb{R}^m \to \mathbb{R}$.*

1. If the supply rate $\xi\big(k, x(k), u(k)\big)$ satisfies the bounded condition (A.2), then $\big|\xi\big(k, x(k), u(k)\big)\big| \to 0$ as $k \to +\infty$.

2. If the supply rate $\xi\big(k, x(k), u(k)\big)$ is $\mathcal{K}\mathcal{L}$-bounded, then the boundedness of (A.2) holds.

Proof (1) The boundedness of (A.2) $\Rightarrow \lim\limits_{k \to \infty} \big|\xi\big(k, x(k), u(k)\big)\big| = 0$:

Assume, on the contrary, that $\lim\limits_{k \to \infty} \big|\xi\big(k, x(k), u(k)\big)\big| = \delta > 0$, i.e. for any $k > k_0$ there is always a small $\nu(k) > 0$ such that $\big|\xi\big(k, x(k), u(k)\big)\big| \geqslant \delta + \nu(k)$. Therefore, $\lim\limits_{\kappa \to +\infty} \sum\limits_{k=k_0+1}^{\kappa} \big|\xi\big(k, x(k), u(k)\big)\big| = +\infty$. This means the bounded condition (A.2) is not true.

(2) GDC $\Rightarrow$ the boundedness (A.2): In what follows, we denote $\xi_{(k)} := \xi\big(k, x(k), u(k)\big)$, and thus, $\xi_{(k_0)} := \xi\big(k_0, x(k_0), u(k_0)\big)$ for conciseness.

The $\mathcal{K}\mathcal{L}$ function $\alpha\big(\big|\xi_{(k_0)}\big|, k-k_0\big)$ is expressed as a product of a class K function and a small real number in this proof. Consider an indexed set of real numbers for each $\alpha(.,.)$ and k_0

$$\mathscr{E} = \{\varepsilon_k \in \mathbb{R}^+ \big| k \in \mathbb{N}, \ k > k_0 : \varepsilon_k \leqslant 1 \wedge \lim\limits_{k \to +\infty} \prod\limits_{i=1+k_0}^{k} \varepsilon_i = 0\}.$$

The value of α at the time step k can be related to its value at the time step $k - 1$, as follows:

$$\alpha\big(\big|\xi_{(k_0)}\big|, k - k_0\big) \leqslant \varepsilon_k \times \alpha\big(\big|\xi_{(k_0)}\big|, k - 1 - k_0\big),$$

for all $k > k_0$, and to its value at the initial time step k_0:

$$\alpha\big(\big|\xi_{(k_0)}\big|, k - k_0\big) \leqslant \alpha\big(\big|\xi_{(k_0)}\big|, 0\big) \times \prod\limits_{i=1+k_0}^{k} \varepsilon_i. \tag{A.5}$$

It is noted here that, $\varepsilon_k \leqslant 1$ is due to the decreasing (not strictly) property of the $\mathscr{K}\mathscr{L}$ function $\alpha(s, k)$ for a fixed s. And $\prod\limits_{i=1+k_0}^{k} \varepsilon_i \to 0$ as $k \to +\infty$ due to $\alpha(., k)$ also goes to zero as $k \to +\infty$. Then, it follows from (A.4) that

$$\sum_{k=k_0+1}^{\kappa} \left|\xi\big(k, x(k), u(k)\big)\right| \leqslant \alpha\big(\big|\xi_{(k_0)}\big|, 0\big) \sum_{k=k_0+1}^{\kappa} \Big(\prod_{j=1+k_0}^{k} \varepsilon_j \Big). \tag{A.6}$$

For $\varepsilon_m > \max_{k=k_0+1}^{\kappa} \varepsilon_k > 1$, denote $\varepsilon_{(k)} := \dfrac{\varepsilon_k}{\varepsilon_m}$. It is obviously that $0 < \varepsilon_{(k)} < 1$. (A.6) is then equivalent to

$$\sum_{k=k_0+1}^{\kappa} \left|\xi\big(k, x(k), u(k)\big)\right| \leqslant \varepsilon_m \, \alpha\big(\big|\xi_{(k_0)}\big|, 0\big) \sum_{k=k_0+1}^{\kappa} \Big(\prod_{j=1+k_0}^{k} \varepsilon_{(j)} \Big). \tag{A.7}$$

Applying the sum of consecutive powers, we get

$$\sum_{k=k_0+1}^{\kappa} \left|\xi\big(k, x(k), u(k)\big)\right| \leqslant \varepsilon_m \, \alpha\big(\big|\xi_{(k_0)}\big|, 0\big) \frac{1 - \varepsilon_{(m)}^{\kappa+1}}{1 - \varepsilon_{(m)}}, \tag{A.8}$$

where $\varepsilon_{(m)} := \max_{k=k_0+1}^{\kappa} \varepsilon_{(k)}$. The boundedness of (A.2) is then obtained with $\theta = \varepsilon_m \, \alpha\big(\big|\xi_{(k_0)}\big|, 0\big) \dfrac{1}{1 - \varepsilon_{(m)}}$. The proof is complete ∎

In other contexts, the GDC can be defined in association with the dissipation inequality for a more general nonlinear system of the form

$$\mathscr{S} : x(k + 1) = f(x(k), u(k)), \tag{A.9}$$

where $x \in \mathbb{X} \subset \mathbb{R}^n$, $u \in \mathbb{U} \subset \mathbb{R}^m$ are the state and control vectors, respectively; $\mathbb{X}$ is compact, $0 \in \mathbb{X}$, $0 \in \mathbb{U}$, and $f(0, 0) = 0$; f is not necessarily continuous. The controlled system with MPC is usually discontinuous in x, even when the open-loop system is continuous. The function $V : \mathbb{Z} \times \mathbb{R}^n \to \mathbb{R}_0^+$ in the GDC method is generally piecewise continuous, whereas smooth or locally Lipschitz continuous Lyapunov functions are usually considered for discontinuous dynamical systems in previous works. The assumption on the locally Lipschitz continuity is not made, but the bounds of the form $\underline{\alpha}\big(\|x(k)\|\big) \leqslant V\big(k, x(k)\big) \leqslant \overline{\alpha}\big(\|x(k)\|\big)$ are employed in the stability conditions instead.

Definition A.4 The controlled system $\mathscr{S}$ (A.9) is said to be *GDC stable* around the zero equilibrium with respect to the supply rate $\xi\big(k, x(k), u(k)\big)$ and the real-valued, non-negative and radially unbounded storage function $V(k, x(k))$ if the following conditions hold for all $k > 0$ with some control sequences $u(k) \in \mathbb{R}^m$, irrespective of the initial state $x(0)$:

$$\underline{\alpha}(\|x(k)\|) \leqslant V(k, x(k)) \leqslant \overline{\alpha}(\|x(k)\|),$$

$$V(k, x(k)) - \tau V(k-1, x(k-1)) \leqslant |\xi(k, x(k), u(k))|, \quad 0 < \tau < 1, \text{ and}$$

$$|\xi(k, x(k), u(k))| \leqslant \alpha\left(\left|\xi(0, x(0), u(0))\right|, k\right).$$

for some $\mathcal{K}_\infty$ functions $\underline{\alpha}(.)$ and $\overline{\alpha}(.)$, and some $\mathcal{K}\mathcal{L}$ function $\underline{\alpha}(.)$.

Alternatively, the GDC stability can be defined with a separated dissipation inequality, as follows:

Definition A.5 The controlled system $\mathscr{S}$ (A.9) is said to be *GDC stable* around the zero equilibrium with respect to the supply rate $\xi(k, x(k), u(k))$ and the real-valued, non-negative, and radially unbounded storage function $V(k, x(k))$ if the following conditions hold for all $u(k) \in \mathbb{R}^m$ and all $k > 0$, irrespective of the initial state $x(0)$:

$$\underline{\alpha}(\|x(k)\|) \leqslant V(k, x(k)) \leqslant \overline{\alpha}(\|x(k)\|),$$

$$V(k, x(k)) - \tau V(k-1, x(k-1)) \leqslant |\xi(k, x(k), u(k))|, \quad 0 < \tau < 1$$

for some $\mathcal{K}_\infty$ functions $\underline{\alpha}(.)$ and $\overline{\alpha}(.)$, and there exist some control sequences $u(k) \in \mathbb{U} \subset \mathbb{R}^m$, such that the following inequality is fulfilled for all $k > 0$:

$$|\xi(k, x(k), u(k))| \leqslant \alpha\left(\left|\xi(0, x(0), u(0))\right|, k\right),$$

for some $\mathcal{K}\mathcal{L}$ function $\underline{\alpha}(.)$.

The GDC will be applied to the problem of designing the closed-form control law or in the optimisation-based control algorithms such as the model predictive control for $\mathscr{S}$. In the next section, the input-to-power-and-state stabilisation with the GDC is introduced. The GDC method is different to the previous constructive method by considering the controlled system as a single system instead of as an interconnection of two open-loop systems. As a result, controlled system $\mathscr{S}$ (A.9) can be simply said to satisfy some GDCs in Definition A.4; i.e., the dissipation inequality is assumed an integrated part of the GDC.

A.2 Input-to-Power-and-State Stabilisation

In the GDC method, an input-to-power-and-state stable (IpSS) closed-loop system will be obtained when either a memoryless casual control law or a control sequence from an optimisation-based control algorithm is applied such that the GDC inequalities are fulfilled. The IpSS is firstly defined below.

Definition A.6 The controlled system $\mathscr{S}$ (A.1) is said to be IpSS stabilised if there are two functions α_i of class $\mathscr{KL}$, $i = \{0, 1\}$, a finite initial supply rate $\xi_{(k_0)}$ and a function γ of class $\mathscr{K}$, such that for each initial state $x(k_0) = x_{k_0}$, the following inequality is satisfied for all $k > k_0$:

$$\|x(k, x_{k_0}, d)\| \leqslant \alpha_0(\|x_{k_0}\|, k - k_0) + \alpha_1(|\xi_{(k_0)}|, k - k_0) + \gamma(\|d\|_\infty), \quad\quad \text{(A.10)}$$

with some admissible control sequences $\{u(k) \in \mathbb{U}\}$.

It is noted here that $\|x(k)\|$ may diverge more aggressively during certain time intervals when the term $\alpha_1(|\xi_{(k_0)}|, k - k_0)$ is additionally included in (A.10).

The stability condition is stated in the next theorem. For a real-valued non-negative function $V(k, x)$, $V : \mathbb{Z} \times \mathbb{R}^n \to \mathbb{R}_0^+$, denote

$$\Delta_\tau V\big(k, x(k), x(k-1)\big) := V\big(k, x(k)\big) - \tau V\big(k, x(k-1)\big), \quad 0 < \tau < 1. \quad\quad \text{(A.11)}$$

Theorem A.1 *Consider the nominal system $\mathscr{S}$ (A.1) with vanishing $d(k)$, and a real-valued piecewise-continuous supply rate $\xi\big(k, x(k), u(k)\big)$, $\xi : \mathbb{Z} \times \mathbb{R}^n \times \mathbb{R}^m \to \mathbb{R}$. Let $\tau \in \mathbb{R}^+$, $\tau < 1$. Suppose that the bounded condition (A.2) on $\xi\big(k, x(k), u(k)\big)$ holds true and there are two $\mathscr{K}_\infty$ functions $\underline{\alpha}(\|x\|)$, $\overline{\alpha}(\|x\|)$ and a real-valued, piecewise-continuous, non-negative, and radially unbounded (in x), function $V\big(k, x(k)\big)$, $V : \mathbb{Z} \times \mathbb{R}^n \to \mathbb{R}_0^+$, $V\big(k_0, x(k_0)\big)$ is finite, such that for each $k_0 \geqslant 0$ and each $x(k_0) \in \mathbb{X}$, the following conditions hold for all $k > k_0$:*

1. $\underline{\alpha}(\|x(k)\|) \leqslant V\big(k, x(k)\big) \leqslant \overline{\alpha}(\|x(k)\|)$,

2. $\Delta_\tau V\big(k, x(k), x(k-1)\big) \leqslant \big|\xi\big(k, x(k), u(k)\big)\big|$,

3. $V\big(k, x(k-1)\big) \leqslant V\big(k-1, x(k-1)\big)$;

with some admissible control sequences $\{u(k) \in \mathbb{U}\}$.

Then $x(k)$ of $\mathscr{S}$ (A.1) is quadratically attractive, i.e. $\|x(k)\| \to 0$ as $k \to 0$.

Proof The evolution of $V\big(k, x(k)\big)$—from the conditions 2 and 3 in Theorem A.1, the following inequality is obtained for all $k > k_0$:

$$
\begin{aligned}
&V\big(k, x(k)\big) \\
&\leqslant \tau V\big(k, x(k-1)\big) + \big|\xi\big(k, x(k), u(k)\big)\big| \quad\quad\quad\quad\quad\quad\quad\quad\quad \text{(A.12)}\\
&\leqslant \tau V\big(k-1, x(k-1)\big) + \big|\xi\big(k, x(k), u(k)\big)\big| \\
&\leqslant \tau\big[\tau V\big(k-1, x(k-2)\big) + \big|\xi\big(k-1, x(k-1), u(k-1)\big)\big|\big] + \big|\xi\big(k, x(k), u(k)\big)\big| \\
&\leqslant \tau^2 V\big(k-2, x(k-2)\big) + \big[\tau\big|\xi\big(k-1, x(k-1), u(k-1)\big)\big| + \big|\xi\big(k, x(k), u(k)\big)\big|\big].
\end{aligned}
$$

$$\text{(A.13)}$$

Continuing in this way, we get

$$V\big(k, x(k)\big) \leqslant \tau^{k-k_0} V\big(k_0, x(k_0)\big) + \sum_{i=k_0+1}^{k-k_0-1} \tau^i \big|\xi\big(i, x(k-k_0-i), u(k-k_0-i)\big)\big|. \quad \text{(A.14)}$$

Applying the convolution sum and (1) of Lemma A.1 to the second term on the right-hand side of (A.14), we have both the first and second terms on the right-hand side of (A.14) goes to zero as $k \to +\infty$, due to $0 < \tau < 1$. Accordingly, using $\underline{\alpha}(\|x(k)\|) \leqslant V(k, x(k)) \leqslant \overline{\alpha}(\|x(k)\|)$ in condition 1 in Theorem A.1, it can be concluded that $\|x(k)\| \to 0$ as $k \to 0$. The radially unbounded function $V(x)$ ensures that the stated condition is not restricted to some local conditions. The proof is complete ∎

The next theorem states a sufficient stability condition with the GDC.

Theorem A.2 *Consider the system $\mathscr{S}$ (A.1) and a real-valued piecewise-continuous supply rate $\xi\big(k, x(k), u(k)\big)$, $\xi : \mathbb{Z} \times \mathbb{R}^n \times \mathbb{R}^m \to \mathbb{R}$ with $\xi\big(k, x(k), u(k)\big)$ is finite for each $k > k_0$. Let $\tau \in \mathbb{R}^+, \tau < 1$. Suppose there are two $\mathscr{K}_\infty$ functions $\underline{\alpha}(\|x\|), \overline{\alpha}(\|x\|)$ and a real-valued, piecewise-continuous, non-negative, and radially unbounded (in x), function $V(k, x(k))$, $V : \mathbb{Z} \times \mathbb{R}^n \to \mathbb{R}_0^+$, $V\big(k_0, x(k_0)\big)$ is finite, such that for each $k_0 \geqslant 0$ and each $x(k_0) \in \mathbb{X}$, the following conditions hold for all $k > k_0$:*

1. $\underline{\alpha}\big(\|x(k)\|\big) \leqslant V\big(k, x(k)\big) \leqslant \overline{\alpha}\big(\|x(k)\|\big)$,

2. $\Delta_\tau V\big(k, x(k), x(k-1)\big) \leqslant \big|\xi\big(k, x(k), u(k)\big)\big| + \sigma\, d(k)^T d(k), \quad \sigma \in \mathbb{R}^+,$

3. $\xi\big(k, x(k), u(k)\big)$ is $\mathscr{KL}$ bounded, and

4. $V\big(k, x(k-1)\big) \leqslant V\big(k-1, x(k-1)\big)$;

with some admissible control sequences $\{u(k) \in \mathbb{U}\}$.
Then $\mathscr{S}$ (A.1) is IpSS stabilised and the nominal controlled system is quadratically attractive.

Proof (1) *Attractiveness*: From (2) in Lemma A.1, the boundedness of (A.2) is obtained from the condition 3 in Theorem (A.2). The attractiveness of $\mathscr{S}$ is thus obtained as a redirect result of Theorem A.1 when $d = 0$.
(2) *IbSS stabilisability when $d(k) \neq 0$*: From the conditions 2 and 4 in Theorem A.2, the following inequality is obtained for all $k > k_0$:

$$\begin{aligned}
V\big(k, x(k)\big) &\leqslant \tau V\big(k, x(k-1)\big) + \big|\xi\big(k, x(k), u(k)\big)\big| + \sigma d(k)^T d(k) \\
&\leqslant \tau V\big(k-1, x(k-1)\big) + \big|\xi\big(k, x(k), u(k)\big)\big| + \sigma d(k)^T d(k) \\
&\leqslant \tau\big[\tau V\big(k-1, x(k-2)\big) + \big|\xi\big(k-1, x(k-1), u(k-1)\big)\big| \\
&\quad + \sigma d(k-1)^T d(k-1)\big] + \big|\xi\big(k, x(k), u(k)\big)\big| \sigma d(k)^T d(k) \\
&\leqslant \tau^2 V\big(k-2, x(k-2)\big) + \big[\tau \big|\xi\big(k-1, x(k-1), u(k-1)\big)\big| \\
&\quad + \big|\xi\big(k, x(k), u(k)\big)\big|\big] + \sigma\big[\tau d(k-1)^T d(k-1) + d(k)^T d(k)\big].
\end{aligned}$$

Continuing in this way, and using $d^T(k)d(k) \leqslant \theta$ for each k and, from condition 3 of Theorem A.2, that

$$\left| \xi(k, x(k), u(k)) \right| \leqslant \alpha \left(\left| \xi(k_0, x(k_0), u(k_0)) \right|, k - k_0 \right),$$

we get

$$V(k, x(k)) \leqslant \tau^{k-k_0} V(k_0, x(k_0)) + \sum_{i=0}^{k-k_0-1} \tau^i \alpha \left(\left| \xi(k_0, x(k_0), u(k_0)) \right|, k - k_0 - i \right)$$

$$+ \sigma\theta \sum_{i=0}^{k-k_0-1} \tau^i. \tag{A.15}$$

Applying equality (A.5) to the second term on the right-hand side of (A.15), we have

$$V(k, x(k)) \leqslant \tau^{k-k_0} V(k_0, x(k_0)) + \alpha \left(\left| \xi(k_0, x(k_0), u(k_0)) \right|, 0 \right) \times \sum_{i=0}^{k-k_0-1} \left(\tau^i \prod_{j=1+k_0}^{k-i} \varepsilon_j \right)$$

$$+ \sigma\theta \sum_{i=0}^{k-k_0-1} \tau^i, \quad \text{where } \varepsilon_j \in \mathscr{E}$$

$$= \tau^{k-k_0} V(k_0, x(k_0)) + \alpha \left(|\xi_{k_0}|, 0 \right) \sum_{i=0}^{k-k_0-1} \left(\tau^i \prod_{j=1+k_0}^{k-i} \varepsilon_j \right) + \sigma\theta \sum_{i=0}^{k-k_0-1} \tau^i. \tag{A.16}$$

<u>Case i:</u> $0 < \tau < \varepsilon_k \leqslant 1$ for all $\varepsilon_k \in \mathscr{E}$.

Denote $\tau_{(k)}^i := \dfrac{\tau^i}{\prod_{j=k-i+1}^{k} \varepsilon_j}$, it follows from (A.16) that

$$V(k, x(k)) \leqslant \tau^{k-k_0} V(k_0, x(k_0)) + \alpha \left(|\xi_{k_0}|, 0 \right) \prod_{j=1+k_0}^{k} \varepsilon_j \sum_{i=0}^{k-k_0-1} \tau_{(k)}^i + \sigma\theta \sum_{i=0}^{k-k_0-1} \tau^i. \tag{A.17}$$

By the sum of powers, we have

$$V(k, x(k)) \leqslant \tau^{k-k_0} V(k_0, x(k_0)) + \alpha \left(|\xi_{k_0}|, 0 \right) \frac{1 - \tau_m^{k-k_0}}{1 - \tau_m} \prod_{j=1+k_0}^{k} \varepsilon_j + \sigma\theta \frac{1 - \tau^{k-k_0}}{1 - \tau}, \tag{A.18}$$

for all $k > k_0$, where $\tau_m := \max_{1 \leqslant i \leqslant k-1} \tau_{(i)}, 0 < \tau_m < 1$.

Case ii: $1 > \tau > \varepsilon_k$ for all $\varepsilon_k \in \mathscr{E}$.

Denote $\varepsilon_{(k)} := \dfrac{\varepsilon_k}{\tau}$, it follows from (A.16) and the sum of powers that

$$V\big(k, x(k)\big) \leqslant \tau^{k-k_0} V\big(k_0, x(k_0)\big) + \tau^{k-k_0-1}\alpha\big(|\xi_{k_0}|, 0\big)\frac{1 - \varepsilon_m^{k-k_0}}{1 - \varepsilon_m} + \sigma\theta\frac{1 - \tau^{k-k_0}}{1 - \tau},$$

$$(A.19)$$

for all $k > k_0$, where $\varepsilon_m := \max_{1 \leqslant i \leqslant k-1} \varepsilon_{(i)}, 0 < \varepsilon_m < 1$.

Case iii: $1 > \tau > \varepsilon_i \wedge 1 \geqslant \varepsilon_j > \tau > 0$ for some $\varepsilon_i \in \mathscr{E}$ and $\varepsilon_j \in \mathscr{E}$.

It follows from (A.18) and (A.19) that

$$V\big(k, x(k)\big) \leqslant \tau^{k-k_0} V\big(k_0, x(k_0)\big) + \sigma\theta\frac{1 - \tau^{k-k_0}}{1 - \tau} \tag{A.20}$$

$$+\alpha\big(|\xi_{k_0}|, 0\big) \times \max\left(\frac{1 - \tau_m^{k-k_0}}{1 - \tau_m}\prod_{j=1+k_0}^{k}\varepsilon_j, \ \tau^{k-k_0-1}\frac{1 - \varepsilon_m^{k-k_0}}{1 - \varepsilon_m}\right).$$

Now, applying the inequality

$$\underline{\alpha}^{-1}(y_1 + y_2 + y_3) \leqslant \underline{\alpha}^{-1}(2y_1) + \underline{\alpha}^{-1}(4y_2) + \underline{\alpha}^{-1}(4y_3)$$

Sontag [144], we obtain from (A.18) and condition 1

$$\|x(k)\| \leqslant \underline{\alpha}^{-1}\big(2\tau^{k-k_0}\overline{\alpha}(\|x(k_0)\|)\big) + \underline{\alpha}^{-1}\left(4\alpha\big(|\xi_{k_0}|, 0\big)\frac{1 - \tau_m^{k-k_0}}{1 - \tau_m}\prod_{j=1+k_0}^{k}\varepsilon_j\right)$$

$$+\underline{\alpha}^{-1}\left(4\sigma\frac{1 - \tau^{k-k_0}}{1 - \tau}\theta\right),$$

as well as from (A.19) and condition 1

$$\|x(k)\| \leqslant \underline{\alpha}^{-1}\big(2\tau^{k-k_0}\overline{\alpha}(\|x(k_0)\|)\big) + \underline{\alpha}^{-1}\left(4\tau^{k-k_0-1}\alpha\big(|\xi_{k_0}|, 0\big)\frac{1 - \varepsilon_m^{k-k_0}}{1 - \varepsilon_m}\right)$$

$$+\underline{\alpha}^{-1}\left(4\sigma\frac{1 - \tau^{k-k_0}}{1 - \tau}\theta\right).$$

And further,

$$\tau^{k-k_0} \to 0, \quad \frac{1 - \tau_m^{k-k_0}}{1 - \tau_m}\prod_{j=1+k_0}^{k}\varepsilon_j \to 0, \quad \tau^{k-k_0-1}\frac{1 - \varepsilon_m^{k-k_0}}{1 - \varepsilon_m} \to 0,$$

$$\text{and } \sigma \frac{1 - \tau^{k-k_0}}{1 - \tau} \to \frac{\sigma}{1 - \tau}, \text{ as } k \to +\infty,$$

with $0 < \tau < 1$, $0 < \tau_m < 1$, $\theta < +\infty$ and $\sigma < +\infty$, the following inequality is obtained:

$$\|x(k)\| \leqslant \alpha_1(\|x(k_0), k\|) + \alpha_2(|\xi_{(k_0)}|, k - k_0) + \gamma(\|d\|_\infty), \; k_0 \geqslant 0,$$

where $\alpha_1(s, k) \geqslant \underline{\alpha}^{-1}(2\tau^k \overline{\alpha}(s))$, $\gamma(s) \geqslant \underline{\alpha}^{-1}\left(\frac{\sigma}{1 - \tau}\beta(s)\right)$, and

$$\alpha_2(s, k) \geqslant \underline{\alpha}^{-1}\left(\max\left(4\varepsilon_0\alpha(|\xi_{k_0}|, 0), \frac{\tau^{k-1}}{1 - \varepsilon_m}4\alpha(|\xi_{k_0}|, 0)\right)\right),$$

in which $\varepsilon_0 := \frac{\prod_{j=1+k_0}^{k}\varepsilon_j}{1-\tau_m}$, $\beta(s) = 4s$, α_1 and α_2 are the two class $\mathscr{K}\mathscr{L}$ functions and γ is a class $\mathscr{K}$ function. The IpSS inequality (A.10) is thus obtained. The proof is complete ∎

The condition for IpSS in Theorem A.2 is restated in a corollary below when

$$V(k, x(k)) = x(k)^T P(k) x(k), \; P(k) = P(k)^T \succ 0,$$

$$W(k, x(k-1)) := x(k-1)^T P(k-1) x(k-1), \text{ and}$$

$$\Delta_\tau W(k, x(k), x(k-1)) := V(k, x(k)) - \tau W(k, x(k-1)) \qquad (A.21)$$

are employed in condition 2 in Theorem A.2.

Corollary A.1 *Consider the system $\mathscr{S}$ (A.1) and a real-valued piecewise-continuous supply rate $\xi(k, x(k), u(k))$, $\xi : \mathbb{Z} \times \mathbb{R}^n \times \mathbb{R}^m \to \mathbb{R}$. Let $\tau \in \mathbb{R}^+$ with $\tau < 1$. Suppose there are two real-valued, piecewise-continuous, non-negative, and radially unbounded, functions $V(k, x(k)) := x(k)^T P(k) x(k)$, $P(k) \succ 0$, and $W(k, x(k-1)) = x(k-1)^T P(k-1) x(k-1)$, $P(k-1) \succ 0$, such that for each $x(0) \in \mathbb{X}$ the following conditions hold for all $k > 0$:*

1. $\Delta_\tau W(k, x(k), x(k-1)) \leqslant |\xi(k, x(k), u(k))| + \sigma d(k)^T d(k)$, $\sigma \in \mathbb{R}^+$, and

2. $\xi(k, x(k), u(k))$ is $\mathscr{K}\mathscr{L}$-bounded,

with some admissible control sequences $\{u(k) \in \mathbb{U}\}$.

Then, $\mathscr{S}$ (A.1) is IpSS stabilised.

Proof Since $P(k) \succ 0$, the condition (1) in Theorem A.2 is fulfilled. From the definition of $\Delta_\tau W(k, x(k), x(k-1))$, which is $\Delta_\tau V(k, x(k), x(k-1))$ in Theorem A.2 with $V(k, x(k-1))$ is replaced by $W(k, x(k-1))$, we deem obtain

$W\big(k, x(k-1)\big) = V\big(k-1, x(k-1)\big) = x(k-1)^T P(k-1) x(k-1)$, which is a special case of condition 4 in Theorem A.2 $\big(V\big(k, x(k-1)\big) \leqslant V\big(k-1, x(k-1)\big)\big)$. Condition 1 is thus a special case of condition 2 in Theorem A.2. The proof is then similar to that for Theorem A.2 with the initial time step $k_0 = 0$. ∎

Remark A.1 The storage functions in the GDC method are not the ISS Lyapunov functions since $\big|\xi\big(k, x(k), u(k)\big)\big| \geqslant 0$.

A.3 Stability Analysis

The stabilisation with the GDC is governed by the non-negativeness of $\triangle V(x, k)$ along the trajectories (i.e. $\triangle V(x, k) \geqslant 0$), in which $V(x, k) \geqslant 0$ is a storage function. In this section, we analyse the stability that is obtained from the stabilisation with the GDC in the context of Lyapunov stability, Lagrange stability and asymptotic stability. The GDC provides a type of stability that is similar to the Lyapunov stability starting from a future time instant $k^* > 0$. The GDC also provides a boundedness property that is similar to the Lagrange uniform boundedness, but with a feasible condition. As a result, the convergence property for every initial state within the region of interest with the GDC is different to the asymptotic stability in the Lyapunov sense.

There are two main notions of stability in the control literature, namely Lyapunov's and Lagrange's. Several research works in the systems and control field have been centred around the stability theorems of Lyapunov's methods and Lasalle's invariance principle. There have been recent developments for the computerised and networked control systems that extend the traditional methods with non-monotonic Lyapunov functions. Another cornerstone in the classical control literature is the Lagrange stability. The Lagrange stability had been introduced before the time the Lyapunov's methods were becoming well known. However, the Lagrange stability has not been widely used as it only provides a boundedness property or a uniformly bounded system, see, e.g. [96]. Furthermore, the Lyapunov stability defines the stability of a system around an equilibrium point, in other words stability of the equilibria.

An autonomous system is called asymptotically stable around its equilibrium point at the origin if it satisfies the following two conditions:

1. Given any $\varepsilon > 0$, $\exists \delta_1 > 0$ such that if $\|x(t_0)\| < \delta_1$, then $\|x(t)\| < \varepsilon$ $\forall t > t_0$.

2. $\exists \delta_2 > 0$ such that if $\|x(t_0)\| < \delta_2$, then $x(t) \to 0$ as $t \to \infty$.

The first condition requires that the state trajectory can be confined to an arbitrarily small ball centred at the equilibrium point, and of radius ε, when released from an arbitrary initial condition in a ball of sufficiently small radius δ_1. This is called stability in the sense of Lyapunov. It is possible to have stability in the sense of Lyapunov without having asymptotic stability.

The key conditions in the GDC method consist of the dissipation inequality with a storage function, as in the dissipative system theory [172], of the form

$$V(x_k) - \tau V(x_{k-1}) \leqslant \left| (x_k^T \ u_k^T) N (x_k^T \ u_k^T)^T \right|, \ V(x_k) \geqslant 0,$$

plus a dissipation-based inequality of the form

$$\left| (x_k^T \ u_k^T) N (x_k^T \ u_k^T)^T \right| \leqslant \gamma \left| (x_{k-1}^T \ u_{k-1}^T) N (x_{k-1}^T \ u_{k-1}^T)^T \right|,$$

where $N := \begin{pmatrix} Q & S \\ S^T & R \end{pmatrix}$, $\gamma \in (0, 1)$ and $\tau \in (0, 1)$, $x_k = x(k)$, $u_k = u(k)$.

We have proved that $\|x_k\| \to 0$ as $k \to \infty$ with the two above inequalities if $V(x_k) = x_k^T P x_k$, $P \succ 0$. Furthermore, a more general form of dissipation-based inequality with a $\mathcal{KL}$-bounded supply rate has been employed. For $\triangle V(x_k) := V(x_{k+1}) - \tau V(x_k)$, we always have $\triangle V(x_k) \geqslant 0 \ \forall k \geqslant 0$ and $\triangle V(x_k)$ is decreasing (not necessarily monotonically) along the trajectories in this GDC method. This means $V(x_k)$ may increase during some time intervals. This is significantly different to the traditional approach of Lyapunov methods, in which $\triangle V(x_k) \leqslant 0$ for the exponentially stable discrete-time systems. We will show here that the stabilisation with every $x(0) \in \mathbb{X} \subset R^n$, X is compact, employing the two above conditions will not lead to the Lyapunov stability in its original form. One may claim that the Lyapunov stability will eventually be obtained in some future time $k > k^* > 0$. Formally speaking, however, that line of thought is not quite correct since it is not the Lyapunov stability by definition.

a. General Dissipativity Constraint with Lyapunov Stability

The extension of the traditional Lyapunov stability is not new in the control literature. For example, in [179], 'all regularity assumptions on traditional Lyapunov function are removed', and the property of Lyapunov function 'V along the system trajectories is non-increasing' is replaced with 'V along the system trajectories may increase its value during some proper time intervals'.

In the GDC approach, the storage function acts like a relaxed non-monotonic Lyapunov function, but is different to the above relaxed Lyapunov function for switched systems. Here, we consider the following property: 'V along the system trajectories may increase its value during some initial time intervals and then eventually decrease, but not necessarily monotonically, after a certain time instant'. The storage function $V(x(k))$ is a relaxed non-monotonic Lyapunov function only in the GDC method, since $\triangle_\tau(V) := V(x_k) - \tau V(x_{k-1}) \geqslant 0$, and $\triangle_\tau(V)$ is decreasing, not necessarily monotonically.

The GDC stability in the discrete-time domain can be stated as: given any $\varepsilon > 0$, $\exists \delta_0 > 0$ and $\exists k_s > k_0 \geqslant 0$, such that if $\|x(k_0)\| < \delta_0$, then $\|x(k)\| < \varepsilon \ \forall k > k_s$ (instead of $\forall k > k_0 \geqslant 0$). Also, $\exists \delta_c > 0$ such that if $\|x(k_0)\| < \delta_c$, then $x(k) \to 0$ as $k \to \infty$.

b. General Dissipativity Constraint with Lagrange Uniformly Boundedness

The Lagrange stability provides the boundedness property which is different than the Lyapunov stability. It states that a motion of a dynamical system is bounded if $\exists \varepsilon_L$ such that $\|x(t)\| \leqslant \varepsilon_L \ \forall k > k_0$. A dynamical system is *uniformly bounded* if for every $\delta_L > 0$ and for every t_0, there exists an $\varepsilon_L = \varepsilon_L(\delta_L)$, independent of t_0, such that $\|x(0)\| < \delta_L \Rightarrow \|x(t)\| < \varepsilon_L \ \forall t > t_0$.

This is fundamentally different to the Lyapunov stability since the Lyapunov stability starts with every ε, but not δ_L. It is not difficult to show that the stabilisation with the GDC provides a closed-loop or controlled system that has the motions uniformly bounded, provided that the GDC is recursively feasible for every $x(0) \in \mathbb{X}$. Furthermore, the stabilisation with the GDC also provides the converged motion, i.e. $\|x(k)\| \to 0$ as $k \to \infty$, together with the feasibility condition such that this convergence incurs for every $x(0) \in \mathbb{X}$.

c. GDC Stability and IpSS

The question here is whether the stabilisation with the GDC will provide a closed-loop or controlled system, that is (i) 'futurely' Lyapunov stable and converged with a quasi-Lyapunov function of $\|x(k)\| \to 0$ as $k \to \infty$ or (ii) 'conditionally' Lagrange uniformly bounded and converged. Or we should give a new stability concept, such as *stability in the GDC sense*, or simply *GDC stability*, and the *stabilisation in the GDC sense* method, and the GDC method, as an alternative or not. In [162], we have chosen the latter alternative to respect the original notions and definitions. However, the input-to-state stability (ISS) as stated in [162] will only be obtained with an extra condition on the continuity of $u(k)$, and that has been an assumption. In the same spirit with regard to the original notion of ISS [144, 147], we have also introduced the *input-to-power-and-state stability* (IpSS) (and IpSS stabilisabilty) in [158]. This means the 'ISS stabilisability in the GDC sense' will lead to a closed-loop or controlled system that is IpSS without any assumptions on the continuity of $u(k)$ or $\xi(k, u(k), s(k))$ that may restrict the implementation of the GDC method.

Similar to the results in [147] and [65], we have IpSS $\Rightarrow$ GDC stability, but not reversely. The IpSS is defined locally here to ensure that the existence of a solution is practically feasible, since $V(k, x)$ may increase during certain time intervals when the term $\alpha_2(|\xi_0|, k - 1)$, a second $\mathscr{K}\mathscr{L}$ function, is included in the IpSS inequality.

A.4 Model Predictive Control with GDC

The control $u(k)$ is computed online by the model predictive control algorithm that employs the open-loop model of $\mathscr{S}$ (A.1). The traditional objective function of the following [88] is considered:

$$\mathscr{J}(k) = \|x(k+N)\|_{P_o}^2 + \sum_{\ell=1}^{N} \|x(k+\ell)\|_{W_x}^2 + \|u(k+\ell-1)\|_{W_u}^2,$$

where W_x, W_u are weighting matrices, and N is the predictive (and control) horizon. The weighting coefficients in W_x, W_u are tuning parameters. The term $\|x(k+N)\|_{P_o}^2$ is called 'terminal cost'. The notation $\|x\|_{W_x}$ is the weighted ℓ_2−norm of x, $W_x \geqslant 0$. The current state vector $x(k)$ is assumed to be known. The optimisation problem of minimising $\mathscr{J}(k)$ subject to the open-loop model of $\mathscr{S}$, the state and control constraints $x \in \mathbb{X}$ and $u \in \mathbb{U}$, respectively, and the GDC $|\xi(k, x(k), u(k))| \leqslant \alpha(|\xi(0, x(0), u(0))|, k)$, formulated in the following:

$$\min_{\{\hat{u}\}} \mathscr{J}(k) \tag{A.22}$$
$$\text{subject to} \quad (A.1), \ x \in \mathbb{X}, \ u \in \mathbb{U}, \ \text{and the GDC},$$

is then solved for the optimising vector sequence $\{\hat{u}\}$ which consists of N elements of $u^*(k+\ell)$, $\ell = 0, 1, \ldots, N-1$. Only the first element $u^*(k)$ of the sequence is applied to Σ. This rolling process is repeated at the next time step and continues thereon. The GDC has the form of $u^*(k) \in \mathbb{V} \subset \mathbb{R}$, in which $0 \in \mathbb{V}$. If P_o is from the Riccati equation as in the control literature, the solution to (A.22) without the GDC for the linearised system will be identical to that of the LQR problem.

Stability Condition

The closed-loop system stability will be formed by the following conditions: (i) the state convergence; (ii) the GDC feasibility; (iii) the recursive feasibility with invariant sets delineated in the following:

(1) $\mathbb{X}_f$—Terminal constraint set: The state evolution beyond the predictive horizon should belong to a *terminal constraint set*, which is a positively invariant set, denoted as $\mathbb{X}_f$, [93], to guarantee the recursive feasibility of MPC.

The terminal constraint set for linear systems is often chosen as the *maximal output admissible set*, O_∞ [42] of the closed-loop system of the form $x(k+1) = (A + BK_f) x(k)$, where the control law $u = K_f x$ is the optimal controller for the unconstrained infinite horizon LQ problem in the linear system cases. From the maximal output admissible set $\mathbb{X}_f$, the initial feasible set is then computed.

(2) $\mathbb{X}_r$—Initial feasible set: The set $\mathbb{X}_r$ is the initial feasible set w.r.t $\mathbb{X}_f$ for system $x(k+1) = A(k) + Bu(k)$, with constraints $x \in \mathbb{X}$, $u \in \mathbb{U}$ [68], if and only if there exists an admissible control law that will drive the state of the system into $\mathbb{X}_f$ in N steps or less from $\mathbb{X}_r$, while keeping the evolution of x inside $\mathbb{X}$, i.e.

$$\mathbb{X}_r := \{x(k) \in \mathbb{X} \mid \exists \{u(k) \in \mathbb{U}\}_{k=0}^{N-1} : \{x(k) \in \mathbb{X}\}_{k=0}^{N-1} \wedge x(N) \in \mathbb{X}_f\}.$$

The above initial feasible set $\mathbb{X}_r$ is determined by computing backward from the terminal constraint set $\mathbb{X}_f$ using set operations. The representation of $\mathbb{X}_r$ in relation to $\mathbb{X}_f$ using the Minkowski sum for discrete-time systems can be found in [122]. The constrained optimisation of MPC is then recursively feasible (for all time $k \geqslant 0$) if and only if the initial state $x(0)$ belongs to the initial feasible set $\mathbb{X}_r$.

(3) $\mathbb{V}$—One-step admissible control set: Given the current state $x(k) \in \mathbb{X}_r$ and the past control $u(k-1)$, the one-step admissible control set $\mathbb{V}$ is defined as

$$\mathbb{V}(x(k)) := \{u(k) \in \mathbb{U} \mid Ax(k) + Bu(k) \in \mathbb{X}_r\}. \tag{A.23}$$

(4) Recursive feasibility: The condition for assuring the recursive feasibility of the MPC is to have the intersection of $\mathbb{V}_k$ and the GDC ellipsoid denoted as $\mathbb{E}_k$ (governed by the GDC) non-empty, i.e. $\mathbb{V}_k \cap \mathbb{E}_k \neq \emptyset$, and

$$u(k) \in \mathbb{V}_k \cap \mathbb{E}_k. \tag{A.24}$$

The stability condition is then stated below.

Proposition A.1 *Let* $0 < \tau < 1$. *Consider the nominal system* $\mathscr{S}$ *(A.1) and the MPC problem with the predictive horizon N (A.22). Suppose there are two* $\mathscr{H}_\infty$ *functions* $\underline{\alpha}(\cdot)$, $\overline{\alpha}(\cdot)$, *a real-valued non-negative function* $V(k,x)$ *with finite* $V(0, x(0))$ *and a real-valued supply-rate function* $\xi(k, x(k), u(k))$, *such that for each* $x(k_0) \in \mathbb{X}_r$, $\xi(0, x(0), u(0))$ *is finite and the following hold for all* $k > 0$:

1. $\underline{\alpha}(\|x(k)\|) \leqslant V(k, x(k)) \leqslant \overline{\alpha}(\|x(k)\|)$;

2. $V(k, x(k)) - \tau V(k, x(k-1)) \leqslant |\xi(k, x(k), u(k))|$;

3. $|\xi(k, x(k), u(k))| \leqslant \alpha(|\xi(0, x(0), u(0))|, k)$, *where* α *is some* $\mathscr{K}\mathscr{L}$ *function;*

4. $V(k, x(k-1)) \leqslant V(k-1, x(k-1))$;

5. *The one-step admissible control set* $\mathbb{V}_k$ *(A.23) determined from the initial feasible set* $\mathbb{X}_r$ *(A.23) intersects the feasible region* $\mathbb{E}_k$ *governed by the GDC in 3), i.e.* $\mathbb{V}_k \cap \mathbb{E}_k \neq \emptyset$, *and*
$$\boldsymbol{u}(k) \in \mathbb{V}_k \cap \mathbb{E}_k; \tag{A.25}$$

Then, the controlled system $\mathscr{S}$ *is asymptotically stable.* $\square$

Implementation notes: To obtain non-conservative sets $\mathbb{F}_k = \mathbb{V}_k \cap \mathbb{E}_k$ for assuring the recursive feasibility, it is possible to have a sufficiently long predictive horizon N to determine the initial feasible set $\mathbb{X}_r$ off-line. The MPC problem (A.22) can have a shorter predictive horizon for online computation. In the case of $\mathbb{V}_k \cap \mathbb{E}_k = \emptyset$, the supply-rate function will be changed; in other words, $\mathbb{E}_k$ should be adjusted online such that it intersects $\mathbb{V}_k$. This can be performed with the quadratic supply-rate function by having a new set of coefficient matrices Q, S and R at certain time steps $k > 0$.

In the following, the detailed derivations for linear matrix inequalities (LMIs) with the QDC for linear-time-invariant (LTI) systems are provided.

A.5 Linear Matrix Inequalities for Quadratic Dissipativity Constraint

The state feedback design with the QDC is developed in this section. The system in consideration is a single LTI system Σ,

$$\Sigma : x(k+1) = Ax(k) + Bu(k). \tag{A.26}$$

The LMI for an open-loop dissipative system is derived from the dissipation inequality as follows:
The dissipation inequality of the form

$$x_{k+1}^T P x_{k+1} - \tau x_k^T P x_k \leq [x_k^T \; u_k^T] \begin{bmatrix} Q & S \\ S^T & R \end{bmatrix} [x_k^T \; u_k^T]^T \tag{A.27}$$

is equivalent to the following inequality:

$$(*) \begin{bmatrix} A^T PA - \tau P - Q & A^T PB - S \\ * & B^T PB - R \end{bmatrix} \begin{bmatrix} x(k) \\ u(k) \end{bmatrix} \leq 0, \tag{A.28}$$

which is then fulfilled by the following LMI for every $x(k)$ and $u(k)$:

$$\begin{bmatrix} A^T PA - \tau P - Q & A^T PB - S \\ * & B^T PB - R \end{bmatrix} \prec 0. \tag{A.29}$$

Now, with the arrangement of

$$\begin{bmatrix} A^T PA - \tau P - Q & A^T PB - S \\ * & B^T PB - R \end{bmatrix} = \begin{bmatrix} A^T PA & A^T PB \\ * & B^T PB \end{bmatrix} - \begin{bmatrix} \tau P + Q & S \\ * & R \end{bmatrix}$$

$$= \begin{bmatrix} A^T \\ B^T \end{bmatrix} P [A \; B] - \begin{bmatrix} \tau P + Q & S \\ * & R \end{bmatrix},$$

and applying the Schur complement [14], we obtain from (A.29) the following matrix inequality:

$$\begin{bmatrix} P^{-1} & A & B \\ * & \tau P + Q & S \\ * & * & R \end{bmatrix} \succ 0, \tag{A.30}$$

which is equivalent to the following LMI by pre- and post-multiplying with the symmetric matrix $\text{diag}[P,\ I,\ I]$:

$$\begin{bmatrix} P & PA & PB \\ * & \tau P + Q & S \\ * & * & R \end{bmatrix} \succ 0. \tag{A.31}$$

We can rewrite this LMI with the decision variables in bold typeface, as follows:

$$\begin{bmatrix} \boldsymbol{P} & \boldsymbol{PA} & \boldsymbol{PB} \\ * & \tau \boldsymbol{P} + \boldsymbol{Q} & \boldsymbol{S} \\ * & * & \boldsymbol{R} \end{bmatrix} \succ 0. \tag{A.32}$$

This result is well known in the control literature, e.g. [17].

a. State Feedback Synthesis

(i) With the state feedback of $u(k) = Kx(k)$, the dissipation inequality is equivalent to the following inequality:

$$x(k)^T[(A^T + K^T B^T)P(A + BK) - \tau P - M]x(k) \leqslant 0, \tag{A.33}$$

where $M := Q + 2SK + K^T RK$,

which is fulfilled by the following LMI for every $x(k)$:

$$(A^T + K^T B^T)P(A + BK) - \tau P - M \prec 0. \tag{A.34}$$

Then, applying the Schur complement, we obtain the following LMI:

$$\begin{bmatrix} P^{-1} & A + BK \\ * & \tau P + M \end{bmatrix} \succ 0, \tag{A.35}$$

which is equivalent to

$$\begin{bmatrix} P^{-1} & AP^{-1} + BX \\ * & \tau P^{-1} + W \end{bmatrix} \succ 0, \tag{A.36}$$

where $K = XP$ and $W = P^{-1}MP^{-1}$, by multiplying both side with $\text{diag}[I,\ P^{-1}]$, $P \succ 0$.

And with decision variables in bold typeface:

$$\begin{bmatrix} \mathscr{P} \, A\mathscr{P} + BX \\ * \quad \tau\mathscr{P} + W \end{bmatrix} \succ 0, \tag{A.37}$$

where $\mathscr{P} := P^{-1}$.

(ii)The dissipation-based constraint of the form

$$0 \leqslant [x_{k+1}^T \ u_{k+1}^T] N [x_{k+1}^T \ u_{k+1}^T]^T \leqslant \gamma [x_k^T \ u_k^T] N [x_k^T \ u_k^T]^T,$$

where $N = \begin{bmatrix} Q & S \\ S^T & R \end{bmatrix}$, $\gamma \in (0, 1)$, is also obtained for every $x(k)$, if the following matrix inequality holds:

$$\begin{bmatrix} M^{-1} \ A + BK \\ * \quad \gamma M \end{bmatrix} \succ 0. \tag{A.38}$$

The two matrix inequalities (A.36) and (A.38) are then solved for the feasible $K = XP^{-1}$ by having the variable transformation of $M = YP$, as delineated in the following.

b. Solving Two Matrix Inequalities

The task is to solve a system of two matrix inequalities in the following:

$$\begin{bmatrix} P^{-1} \ A + BK \\ * \quad \tau P + M \end{bmatrix} \succ 0, \ P \succ 0, \tag{A.39}$$

$$\begin{bmatrix} M^{-1} \ A + BK \\ * \quad \gamma M \end{bmatrix} \succ 0, \ M \succ 0. \tag{A.40}$$

Suppose that $K = XP$ and $M = YP$, then pre- and post-multiplying (A.39) and (A.40) with $\mathsf{diag}[I, \ P^{-1}]$, we obtain the two following matrix inequalities:

$$\begin{bmatrix} P^{-1} \ AP^{-1} + BX \\ * \quad \tau P^{-1} + W \end{bmatrix} \succ 0, \ P \succ 0, \ W \succ 0, \ \text{and} \tag{A.41}$$

$$\begin{bmatrix} P^{-1}Y^{-1} \ AP^{-1} + BX \\ * \quad \gamma P^{-1}Y \end{bmatrix} \succ 0, \ Y \succ 0, \tag{A.42}$$

where $M = YP$, $W = P^{-1}MP^{-1}$.

These two matrix inequalities have the decision variables of $\mathscr{P} = P^{-1}$, X, W, Y, in which the second matrix inequality is not linear.

The solution will be obtained in two consecutive steps, as follows:

(1) First, find X, W and P from LMI (A.41) only.

Denote the solution as X_0, W_0 and P_0. Thus, $M_0 = P_0 W P_0$, and $Y_0 = M_0 P_0^{-1}$.

(2) Second, substituting $M = Y_0 P$, $X = X_0$ and $Y = Y_0$ to (A.41) and (A.42), we obtain the following two LMIs in $\mathscr{P}$:

$$\begin{bmatrix} \mathscr{P} & A\mathscr{P} + BX_0 \\ * & \tau\mathscr{P} + \mathscr{P}Y_0 \end{bmatrix} \succ 0, \quad \mathscr{P} \succ 0, \tag{A.43}$$

$$\begin{bmatrix} \mathscr{P}Y_0^{-1} & A\mathscr{P} + BX_0 \\ * & \gamma\mathscr{P}Y_0 \end{bmatrix} \succ 0, \tag{A.44}$$

where $\mathscr{P} := P^{-1}$.

Then, we can recover the state feedback gain $K = X_0 P$ and the matrix $M = Y_0 P$.

A generalised dissipation-based constraint has been introduced in this Appendix—the general dissipativity constraint (GDC). The asymptotically positive realness constraint (APRC) and the quadratic dissipativity constraint (QDC) whose supply functions are quadratic are special cases of the GDC. Neither the Lyapunov stability in its original form, nor the Lagrange uniform boundedness is assured by the GDC. The associated storage function in the GDC method is a relaxed non-monotonic Lyapunov function. The GDC stability and the input-to-power-and-state stability and stabilisation (IpSS) have been defined, and the corresponding sufficient condition for use with the MPC has been stated.

References

1. Åstrom, K.J., M. Blanke, A. Isidori, W. Schaufelgerger, and R. Sanz. 2001. *Control of Complex Systems*. Berlin: Springer.
2. Åstrom, K.J., and P.R. Kumar. 2014. Control: A perspective. *Automatica* 50 (1): 3–43.
3. Åstrom, K.J., and R.M. Murray. 2010. *Feedback Systems: An Introduction for Scientists and Engineers*. Princeton: Princeton University Press.
4. Allgöwer F., R. Findeisen, and L.T. Biegler. 2007. *Assessment and Future Directions of Nonlinear Model Predictive Control*. LNCIS, Vol. 358. Berlin: Springer.
5. Allgöwer, F., and A. Zheng. 2000. *Nonlinear Model Predictive Control*. Birkhauser.
6. Angeli, D., and D. Efimov. 2015. A characterization of input-to-state stability for systems with multiple invariant sets. *IEEE Transactions on Automatic Control* 60 (12): 3242–3256.
7. Angeli, D., E.D. Sontag, and Y. Wang. 2000. A characterization of integral input-to-state stability. *IEEE Transactions on Automatic Control* 45: 1082–1097.
8. Antsaklis, P.J., G.K. Befekadu, and V. Gupta. 2013. On reliable stabilization via rectangular dilated LMIs and dissipativity-based certifications. *IEEE Transactions on Automatic Control* 58 (3): 792–796.
9. Astolfi, A. 2005. *Nonlinear and Adaptive Control: Tools and Algorithms for the User*. London: Imperial College Press.
10. Barrabasi, A.L., and R. Albert. 1999. Emerging of scaling in random networks. *Science* 286 (5439): 509–512.
11. Bhatti, T.S., A.A.F. Al-Ademi, and N.K. Bansal. 1997. Load frequency control of isolated wind diesel hybrid power systems. *Energy Conversion and Management* 38 (9): 829–837.
12. Blanchini, F. 1999. Set invariant in control. *Automatica* 35: 1747–1767.
13. Boccia, A., L. Grüne, and K. Worthmann. 2014. Stability and feasibility of state constrained MPC without stabilizing terminal constraints. *Systems and Control Letters* 74: 14–21.
14. Boyd, S., L. ElGhaoui, E. Feron, and V. Balakrishnan. 1994. *Linear Matrix Inequalities in System and Control Theory*. SIAM.
15. Boyd, S., and L. Vandenberghe. 2004. *Convex Optimization*. Cambridge: Cambridge University Press.
16. Brocket R.W. 1995. Stabilization of motor networks. In *Proceedings of the 44thIEEE Conference on Decision and Control*, 1484–1488.
17. Brogliato, B., R. Lozano, B. Maschke, and O. Egeland. 2006. *Dissipative Systems Analysis and Control: Theory and Applications*. Berlin: Springer.

© Springer Nature Singapore Pte Ltd. 2018
A. Tri Tran C. and Q. Ha, *A Quadratic Constraint Approach to Model Predictive Control of Interconnected Systems*, Studies in Systems, Decision and Control 148, https://doi.org/10.1007/978-981-10-8409-6

18. Cai, C., and R. Andrew. 2008. Teel. Input output-to-state stability for discrete-time systems. *Automatica* 44: 326–336.
19. Camacho, E.F., and C. Bordons. 2004. *Model Predictive Control*. Berlin: Springer.
20. Camponogara, E., D. Jia, B.H. Krogh, and S. Talukdar. 2002. Distributed model predictive control. *IEEE Control System Magazine* 1: 44–52.
21. Chen, H., and C.W. Scherer. 2006. Moving horizon H_∞ control with performance adaptation for constrained linear systems. *Automatica* 42: 1033–1040.
22. Christofides, P.D., J.F. Davis, N.H. El-Farra, D. Clark, K.R. Harris, and J.N. Gipson. 2007. Smart plant operation: Vision, progress and challenges. *AiChE Journal* 53: 2734–2741.
23. Clarke, D.W., C. Mohtadi, and P.S. Tuffs. 1987. Generalized predictive control: Part I - The basic algorithm. *Automatica* 23 (2): 137–148.
24. Conte C., N.R. Voellmy, M.N. Zeilinger, M. Morari, and C.N. Jones. 2012. Distributed synthesis and control of constrained linear systems. In *Proceedings of the American Control Conference*, 6017–622.
25. Dashkovskiy, S.N., D.V. Efimov, and E.D. Sontag. 2011. Input to state stability and allied system properties. *Automation and Remote Control* 72 (8): 1579–1614.
26. Desoer, C.A., and M.Y. Wu. 1970. Input-output properties of discrete systems. *Journal of the Franklin Institute* 290: 11–24.
27. Dolk, F., and M. Heemels. 2017. Event-triggered control systems under packet losses. *Automatica* 80: 143–155.
28. Donkers, M.C.F., W.P.M.H. Heemels, D. Bernardini, A. Bemporad, and V. Shneer. 2012. Stability analysis of stochastic networked control systems. *Automatica* 48 (5): 917–925.
29. Dörfler, F., M.R. Jovanović, M. Chertkov, and F. Bullo. 2014. Sparsity-promoting optimal wide-area control of power networks. *IEEE Transactions on Power Systems* 29 (5): 2281–2291.
30. Doyle, J., and G. Stein. 1981. Mutlivariable feedback design: Concepts for a classic/modern synthesis. *IEEE Transactions on Automatic Control* 26: 4–16.
31. Duan, Z., Z.-P. Jiang, and L. Huang. 2017. A new decentralised controller design method for a class of strongly interconnected systems. *International Journal of Control* 90 (2): 201–217.
32. Farina, M., G. Ferrari-Trecate, and R. Scattolini. 2010. Distributed moving horizon estimation for linear constrained systems. *IEEE Transactions on Automatic Control* 55 (11): 2462–2475.
33. Fax, J.A., and R.M. Murray. 2004. Information flow and cooperative control of vehicle formations. *IEEE Transactions on Automatic Control* 49 (9): 1465–1476.
34. Ferrari-Trecate, G., L. Galbusera, M.P.E. Marciandi, and R. Scattolini. 2009. Model predictive control schemes for consensus in multi-agent systems with single- and double-integrator dynamics. *IEEE Transactions on Automatic Control* 54 (11): 2560–2572.
35. Findeisen, R., and P. Varutti. 2008. Stabilizing nonlinear predictive control over nondeterministic communication networks. *LNCIS - Nonlinear Model Predictive Control* 384: 167–179.
36. Forni F., and R. Sepulchre. 2013. On differentially dissipative dynamical systems. In *Proceedings of IFAC symposium in nonlinear control systems*, NOLCOS'13, Toulouse, France.
37. Forni, F., and R. Sepulchre. 2014. A differential lyapunov framework for contraction analysis. *IEEE Transactions on Automatic Control* 59 (3): 614–628.
38. Franklin, G.F., J. Da Powell, and A. Emami-Naeini. 2014. *Feedback Control of Dynamic Systems*, 7th edn. Pearson.
39. Fridman, E. 2014. *Introduction to Time-Delay Systems: Analysis and Control*. Switzerland: Birkhäuser.
40. Gahinet, P., and P. Apkarian. 1994. A linear matrix inequality approach to H_∞ control. *International Journal of Robust and Nonlinear Control* 4 (4): 421–448.
41. Geiselhart, R., and F. Wirth. 2015. On maximal gains guaranteeing a small-gain condition. *SIAM Journal on Control and Optimization* 53 (1): 262–286.
42. Gilbert, E.G., and K.T. Tan. 1991. Linear systems with state and control constraints: The theory and application of maximal output admissible sets. *IEEE Transactions on Automatic Control* 36 (9): 1008–1020.

43. Goodwin, G.C., H. Haimovich, D.E. Quevedo, and J.S. Welsh. 2004. A moving horizon approach to networked control system design. *IEEE Transactions on Automatic Control* 49 (9): 1427–1445.
44. Govindaraju, R., K. Lukman, and D.R. Chandra. 2014. Manufacturing execution system design using ISA-95. *Advanced Materials Research* 980: 248–252.
45. Granjal, J., E. Monteiro, and J.S. Silva. 2015. Security for the Internet of Things: A survey of existing protocols and open research issues. *IEEE Communication Surveys and Tutorials* 17 (3): 1294–1312.
46. Gruhn, P., and H. Cheddie. 2006. *Safety Instrumented Systems: Design, Analysis, and Justification*, 2nd edn. ISA, 2006.
47. Grüne, L. 2002. Input-to-state dynamical stability and its Lyapunov function characterization. *IEEE Transactions on Automatic Control* 47 (9): 1499–1504.
48. Gu, K., and V.L. Kharitonov. 2003. Stability of Time-Delay Systems.
49. Gubbia, J., R. Buyyab, S. Marusic, and M. Palaniswami. 2013. Internet of Things (IoT): A vision, architectural elements, and future directions. *Future Generation Computer Systems* 29: 1645–1660.
50. Gupta, V., A.F. Dana, J.P. Hespanha, R.M. Murray, and B. Hassibi. 2009. Data transmission over networks for estimation and control. *IEEE Transactions on Automatic Control* 54 (8): 1807–1819.
51. Gupta V., B. Sinopoli, S. Adlakha, A. Goldsmith, and R. Murray. 2006. Receding horizon networked control. In *44th Annual Allerton Conference on Communication, Control, and Computing*, 169–176.
52. Haddad, W.M., and D.S. Bernstein. 1991. Robust stabilization with positive real uncertainty: Beyond the small gain theorem. *Systems and Control Letters* 17: 191–208.
53. Haddad, W.M., and V. Chellaboina. 2008. *Nonlinear Dynamical Systems and Control: A Lyapunov-Based Approach*. Princeton: Princeton University Press.
54. Haddad, W.M., E.G. Collins, and D.S. Bernstein. 1993. Robust stability analysis using the small gain, circle, positivity, and Popov theorems: a comparative study. *IEEE Transactions on Automatic Control* 1 (4): 290–293.
55. Halevi, Y., and A. Ray. 1988. Integrated communication and control systems: Part I and II. *Journal of Dynamical Systems, Measurement and Control* 10: 367–381.
56. Hassibi A., S.P. Boyd, and J.P. How. 1999. Control of asynchronous dynamical systems with rate constraint on events. In *Proceedings of the 48thIEEE Conference on Decision and Control*, 1345–1351.
57. Hermans, R.M., A. Jokic, M. Lazar, A. Alessio, P.J. van den Bosch, I.A. Hiskens, and A. Bemporad. 2013. Assessment of non-centralised model predictive control techniques for electrical power networks. *International Journal of Control* 85 (8): 1162–1177.
58. Hespanha, J.P., and A. Mesquita. 2013. Networked control systems: Estimation and control over lossy networks. In *Encyclopedia of Systems and Control*, ed. T. Samad, and J. Baillieul. Berlin: Springer.
59. Hill, D.L., and P.J. Moylan. 1976. The stability of nonlinear dissipative systems. *IEEE Transactions on Automatic Control* 21 (3): 708–711.
60. Hines, G.H., M. Arcak, and A.K. Packard. 2011. Equilibrium-independent passivity: A new definition and numerical certification. *Automatica* 47 (9): 1949–1956.
61. Hokayem, P., D. Chatterjee, F.A. Ramponi, and J. Lygeros. 2012. Stable networked control systems with bounded control authority. *IEEE Transactions on Automatic Control* 57 (12): 3153–3157.
62. Isidori, A. 1999. *Nonlinear Control Systems II*. Berlin: Springer.
63. Jiang, Z.P., A.R. Teel, and L. Praly. 1994. Small gain theorem for ISS systems and applications. *Journal of Mathematics of Control Signals Systems* 7: 95–120.
64. Jiang, Z.P., and Y. Wang. 1997. Ouput-to-state stability and detectability of nonlinear systems. *Systems and Control Letters* 29: 279–290.
65. Jiang, Z.P., and Y. Wang. 2001. Input-to-state stability for discrete-time nonlinear systems. *Automatica* 37: 875–869.

66. Karafyllis, I., M. Malisoff, F. Mazenc, and P. Pierdomenico (eds.). 2016. *Recent Results on Nonlinear Delay Control Systems: In honor of Miroslav Krstic*. Berlin: Springer.

67. Kellett, C.M., and L. Gruene. 2014. ISS-Lyapunov functions for discontinuous discrete-time systems. *IEEE Transactions on Automatic Control* 59 (11): 3098–3103.

68. Kerrigan E.C., and J.M. Maciejowski. 2000. Invariant sets for constrained nonlinear discrete-time systems with application to feasibility in model predictive control. In *Proceedings of the 39th IEEE Conference on Decision and Control*, 4951–4956. Sydney, Australia.

69. Keviczky, T., F. Borrelli, K. Fregene, D. Godbole, and G.J. Balas. 2008. Decentralized receding horizon control and coordination of autonomous vehicle formations. *IEEE Transactions on Control Systems Technology* 16 (1): 19–33.

70. Khalil, H.K. 2002. *Nonlinear Systems*. Pearson Education.

71. Koeln, J.P., and A.G. Alleyne. 2017. Stability of decentralized model predictive control of graph-based power flow systems via passivity. *Automatica* 82: 29–34.

72. Kojima, C., and K. Takaba. 2005. A generalized Lyapunov stability theorem for discrete-time systems based on quadratic difference forms. In *Proceedings of the 44thConference on Decision and Control*.

73. Kolmanovsky, I., and E. Gilbert. 1998. Theory and computation of disturbance invariant sets for discrete-time linear systems. *Mathematical Problems in Engineering: Theory - Methods and Applications* 4: 317–367.

74. Kothare, S.L.O., and M. Morari. 2000. Contractive model predictive control for constrained nonlinear systems. *IEEE Transactions on Automatic Control* 45 (6): 1053–1071.

75. Kouvaritakis, B., and M. Cannon. 2001. *Nonlinear Predictive Control*. IEE Control Engineering Series, Vol. 61. Berlin: Springer.

76. Krichman, M., E.D. Sontag, and Y. Wang. 2001. Input-output-to-state stability. *SIAM Journal of Control and Optimization* 39 (6): 1874–1928.

77. Kwon, O.M., M.-J. Park, S.-M. Lee, J.H. Park, and E.-J. Cha. 2013. Stability for neural networks with time-varying delays via some new approaches. *IEEE Transactions on Neural Networks and Learning Systems* 24 (2): 181–193.

78. Langport C., V. Gupta, and R.M. Murray. 2006. Distributed control over failing channels. *LNCIS - Networked Embedded Sensing and Control*, 325–342.

79. Lazar, M., and R.H. Gielen. 2012. On parameterized dissipation inequalities and receding horizon robust control. *International Journal of Robust and Nonlinear Control* 22 (12): 1314–1329.

80. Leigh J.R. 2004. *Control Theory*, 2nd edn. IEE Control Engineering Series, Vol. 64.

81. Li, C., and G. Chen. 2004. Synchronization in general complex dynamical networks with coupling delays. *Physica A: Statistical Mechanics and its Applications* 343: 263–278.

82. Liu, B., Y. Xia, M. Mahmoud, H. Wu, and S. Cui. 2012. New predictive control scheme for networked control systems. *Journal of Circuits, Systems, and Signal Processing* 31 (2): 945–966.

83. Liu, J., D.M. de la Pena, and P.D. Christophides. 2009. Distributed model predictive control of nonlinear processes. *AIChE Journal* 55 (5): 1171–1184.

84. Liu, K., E. Fridman, and K. Johansson. 2015. Networked control with stochastic scheduling. *IEEE Transactions on Automatic Control* 60 (11): 3071–3076.

85. Liu T., Z-P. Jiang, and D.J. Hill. 2014. *Nonlinear Control of Dynamic Networks*. CRC Press - Taylor and Francis Group.

86. Lunze, J. 1992. *Feedback Control of Large Scale Systems*. New Jersey: Prentice Hall.

87. Lunze, J. 2014. *Control Theory of Digitally Networked Dynamic Systems*. Heidelberg: Springer.

88. Maciejowski, J.M. 2002. *Predictive Control With Constraints*. New Jersey: Prentice Hall.

89. Maestre, J.M., and R.R. Negenborn. 2013. *Distributed Model Predictive Control - Made Easy*. Berlin: Springer Science & Business Media.

90. Magni, L., D.M. Raimondo, and F. Allgöwer. 2009. *Nonlinear Model Predictive Control: Towards New Challenging Applications*. LNCIS, Vol. 384, Berlin: Springer.

91. Mahmoud, M.S. 2000. *Robust Control and Filtering for Time-Delay Systems*. Marcel Dekker.

92. Mahmoud, M.S. 2014. *Control and Estimation Methods over Communication Networks*. UK: Springer.
93. Mayne, D.Q., J.B. Rawlings, C.V. Rao, and P.O.M. Scokaert. 2000. Constraint predictive control: Stability and optimality. *Automatica* 36: 789–814.
94. Mazenc, F., and M. Malisoff. 2017. Stabilization of nonlinear time-varying systems through a new prediction based approach. *IEEE Transactions on Automatic Control* 62 (6): 2908–2915.
95. Miao, Z., and L. Fan. 2017. Achieving economic operation and secondary frequency regulation simultaneously through local feedback control. *IEEE Transactions on Power Systems* 32 (1): 85–93.
96. Michel A.N., L. Hou, and D. Liu. 2015. *Stability of Dynamical Systems - On the Role of Monotonic and Non-Monotonic Lyapunov Functions,* 2nd edn. Birkhäuser.
97. Miorandi, D., S. Sicari, F. De Pellegrini, and I. Chlamtac. 2012. Internet of Things (IoT): A vision, applications and research challenges. *Ad Hoc Networks* 10: 1497–1516.
98. Mishra, P.K., D. Chatterjee, and A.E. Quevedo. 2018. Sparse and constrained stochastic predictive control for networked systems. *Automatica* 87: 40–51.
99. Montestruque, L.A., and P.J. Antsaklis. 2003. On the model-based control of networked systems. *Automatica* 39: 1837–1843.
100. Murray R.M. 2003. *Control in an Information Rich World: Report of the Panel on Future Directions in Control, Dynamics, and Systems*. SIAM.
101. Newman, M. 2010. *Networks: An Introduction*. Oxford: Oxford University Press.
102. Niculescu, S.I. 2001. *Delay Effects on Stability - A Robust Control Approach*. Berlin: Springer.
103. Nocedal, J., and S.J. Wright. 2006. *Numerical Optimization*, 2nd ed. Berlin: Springer.
104. Olfati-Saber, R., J.A. Fax, and R.M. Murray. 2007. Consensus and cooperation in networked multi-agent systems. *Proceedings of the IEEE* 95 (1): 215–233.
105. Patton M., E. Gross, R. Chinn, S. Forbis, L. Walker, and H. Chen. 2014. Uninvited connections – a study of vulnerable devices on the internet of things. In *Proceedings of the IEEE Joint Intelligence and Security Informatics Conference*, 232–235. the Hague, the Netherlands.
106. Pavlov, A., and L. Marconi. 2008. Lyapunov-based control for switched power converters. In *Proceedings of the 45th IEEE Conference on Decision and Control*, 5263–5268.
107. Pena, D.M., and P.D. Christophides. 2008. Lyapunuov-based model predictive control of nonlinear systems subject to data loss. *IEEE Transactions on Automatic Control* 53 (9): 2076–2089.
108. Peng, C., and T.C. Yang. 2013. Event-triggered communication and h_∞ control co-design for networked control systems. *Automatica* 49 (5): 1326–1332.
109. Perera, C., C.H. Liu, and S. Jayawardena. 2015. The emerging Internet of Things marketplace from an industrial perspective: A survey. *IEEE Transactions on Emerging Topics in Computing* 3 (4): 585–598.
110. Petersen, I.R., and R. Tempo. 2014. Robust control of uncertain systems: Classical results and recent developments. *Automatica* 50: 1315–1335.
111. Petersen, I.R., V.A. Ugrinovskii, and A. Savkin. 2000. *Robust Control Design Using H_∞ Methods*. Berlin: Springer.
112. Phung M.D., M. de La Villefromoy, and Q.P. Ha. 2017. Management of solar energy in microgrids using IoT-based dependable control. In *Proceedings of the 20th International Conference on Electric Machines and Systems (ICEMS)*. Sydney, Australia.
113. Pin, G., and T. Parisini. 2011. Networked predictive control of constrained nonlinear systems: Recursive feasibility and input-to-state stability analysis. *IEEE Transactions on Automatic Control* 56 (1): 72–87.
114. Polderman, J.W., and J.C. Willems. 2007. *Introduction to Mathematical Systems Theory: A Behavioural Approach*. Springer: Springer.
115. Prett, D.M., and C.E. Garcia. 1988. Fundamental Process Control. Butterworths.
116. Primbs, J.A., V. Nevistic, and J.C. Doyle. 2000. A receding horizon generalization of pointwise min-norm controllers. *IEEE Transactions on Automatic Control* 45 (5): 898–909.
117. Mayne, D.Q. 2014. Model predictive control: Recent developments and future promise. *Automatica* 50: 2967–2986.

118. Qin, S.J., and T.A. Badgwell. 2003. A survey of industrial model predictive control technology. *Control Engineering Practice* 11: 733–764.
119. Quevedo, D.E., and D. Nesic. 2011. Input-to-state stability of packetized predictive control over unreliable networks affected by packet-dropouts. *IEEE Transactions on Automatic Control* 56 (2): 370–375.
120. Raff T., C. Ebenbauer, and F. Allgöwer. 2005. Nonlinear model predictive control: A passivity-based approach. In *Proceedings of International Workshop on Assessment and Future Directions of Nonlinear Predictive Control NMPC'05*, 151–162.
121. Rakovic, S.V., and R.H. Gielen. 2014. Positively invariant families of sets for interconnected and time-delay discrete-time systems. *SIAM Journal on Control and Optimization* 54 (4): 2261–2283.
122. Rakovic, S.V., E.C. Kerrigan, D.Q. Mayne, and K.I. Kouramas. 2007. Optimized robust control invariance for linear discrete-time systems: Theoretical foundations. *Automatica* 43: 831–841.
123. Rao, C.V., J.B. Rawlings, and J.H. Lee. 2001. Constrained linear state estimation - A moving horizon approach. *Automatica* 37 (10): 1619–1628.
124. Rawlings, J., and D. Mayne. 2009. *Model Predictive Control: Theory and Design*. Wisconsin: Nob Hill Publishing.
125. Rawlings, J.B., and B.T. Stewart. 2008. Coordinating multiple optimization-based controllers: New opportunities and challenges. *Journal of Process Control* 18: 839–845.
126. Robert, D.G. and D.J. Stilwell. 2005. Control of autonomous vehicle platoon with a switch communication network. In *Proceedings of IEEE American Control Conference (ACC'05)*, 4333–4338.
127. Roshany-Yamchi, S., M. Cychowski, R.R. Negenborn, B. De Schutter, K. Delaney, and J. Connell. 2013. Kalman filter-based distributed predictive control of large-scale multi-rate systems: Application to power networks. *IEEE Transactions on Control Systems Technology* 21 (1): 27–39.
128. Rossiter, J.A. 2003. *Model-Based Predictive Control-A Practical Approach*. Boca Raton: CRC Press.
129. Sanders, S.R., and G.C. Verghese. 1992. Lyapunov-based control for switched power converters. *IEEE Transactions on Power Electronics* 7 (1): 17–24.
130. Sastry, S. 1999. *Nonlinear systems: analysis, stability, and control*. Berlin: Springer.
131. Scattolini, R. 2009. Architectures for distributed and hierarchical model predictive control - A Review. *Journal of Process Control* 19: 723–731.
132. Schenato, L., M. Franceschetti, K. Poolla, and S. Sastry. 2007. Foundations of control and estimation over lossy networks. *Proceedings of the IEEE* 95 (1): 164–187.
133. Scherer, C., P. Gahinet, and M. Chilali. 1997. Multi-objective output-feedback control via LMI optimization. *IEEE Transactions on Automatic Control* 42 (7): 896–911.
134. Scherer, C. 2011. Dissipation-based controller synthesis: Successes and challenges. In *Plenary Lecture at the 18th IFAC World Congress*. Milano, Italy.
135. Scherer, C., and J. Mohammadpour (eds.). 2012. *Control of Linear Parameter Varying Systems with Applications*. Berlin: Springer.
136. Schooman, M.L. 2001. *Reliability of Computer Systems and Networks: Fault Tolerance, Analysis and Design*. New Jersey: Wiley-Interscience.
137. Schuler, S., U. Münz, and F. Allgöwer. 2014. Decentralized state feedback control for interconnected systems with application to power systems. *Journal of Process Control* 24 (2): 379–388.
138. Scorletti, G., and G. Duc. 2001. An LMI approach to dencentralized H_∞ control. *International Journal of Control* 74 (3): 211–224.
139. Sepulchre, R., M. Jankovic, and P.V. Kokotovic. 1997. *Constructive Nonlinear Control*. Berlin: Springer.
140. Siljak, D.D. 1991. *Decentralized Control of Complex Systems*. Cambridge: Academic Press.
141. Siljak, D.D. 2006. Dynamic graphs. In *Proceedings of IFAC Conference on Hybrid Systems and Applications*, 544–567.

142. Singh, M.G., and A. Titli. 1978. *Systems: Decomposition*. Franklin Book Company: Optimization and Control.
143. Slotine, J.-J.E., and W. Li. 1991. *Applied Nonlinear Control*. Upper Saddle River: Prentice Hall.
144. Sontag, E.D. 1989. Smooth stabilization implies coprime factorization. *IEEE Transactions on Automatic Control* 34: 435–443.
145. Sontag, E.D. 2007. Input-to-state stability: Basic concepts and results. In *Nonlinear and Optimal Control Theory*, ed. P. Nistri, and G. Stefani, 163–220. Berlin: Springer.
146. Sontag, E.D., and Y. Wang. 1995. On characterizations of the input-to-state stability property. *Systems and Control Letters* 24: 351–359.
147. Sontag, E.D., and Y. Wang. 1996. New characterizations of input to state stability. *IEEE Transactions on Automatic Control* 41 (9): 1283–1294.
148. Tang, P., and C. de Silva. 2005. Stability and optimality of constrained model predictive control with future input buffering in networked control systems. In *Proceedings of IEEE American Control Conference ACC'05*, 1245–1250.
149. Tavassoli, B. 2014. Stability of nonlinear networked control systems over multiple communication links with asynchronous sampling. *IEEE Transactions on Automatic Control* 59 (2): 511–515.
150. Tavassoli, B. 2014. Stability of nonlinear networked control systems over multiple communication links with asynchronous sampling. *IEEE Transactions on Automatic Control* 59 (2): 511–515.
151. Tippett, M.J., and J. Bao. 2013. Distributed model predictive control based on dissipativity. *AIChE Journal* 59 (3): 787–804.
152. Tri Tran. 2013. Dependable model predictive control for reliable constrained systems. In *Proceedings of the 2ndInternational Conference on Control, Automation and Information Science*, 128–133, Nha Trang, Vietnam.
153. Tri Tran., and Q.P. Ha. 2011. Networked control systems with accumulative quadratic constraint. *Electronics Letters* 4 (2): 108–110.
154. Tri Tran., and Q.P. Ha. 2014. Decentralised model predictive control for networks of linear systems with coupling delay. *Journal of Optimization Theory and Applications* 161 (3): 933–950.
155. Tri Tran., and Q.P. Ha. 2014. Dependable control systems with self-recovery constraint. In *Proceedings of the 3rdInternational Conference on Control, Automation and Information Science*, 87–92, Gwangyu, South Korea.
156. Tri Tran., and Q.P. Ha. 2015. Dependable control systems with Internet of Things. *ISA Transactions* 59: 303–313.
157. Tri Tran., and Q.P. Ha. 2018. Perturbed cooperative-state feedback strategy for model predictive networked control of interconnected systems. *ISA Transactions*, [Available. 6 October 2017].
158. Tri Tran., and K-V. Ling, J. Maciejowski. 2014. Closed-loop development for quadratic dissipativity constraint. In *Proceedings of the 3rd International Conference on Control, Automation and Information Science*, 24–29, Gwangju, South Korea.
159. Tri Tran, K-V. Ling, and J. Maciejowski. 2014. Model predictive control via quadratic dissipativity constraint. In *Proceedings of IEEE 53rd Anual Conference on Decision and Control*, 6689–6694, Los Angeles.
160. Tri Tran, K-V. Ling, and J. Maciejowski. 2015. Multiplexed model predictive control of interconnected systems. In *Proceedings of IEEE 54thAnnual Conference on Decision and Control*, 2383–2388, Osaka, Japan.
161. Tri Tran, J. Maciejowski, and K-V. Ling. 2018. A general dissipativity constraint for feedback system design. To appear in *International Journal of Robust and Nonlinear Control*.
162. Tri Tran., and J.M. Maciejowski. 2016. Control with general dissipativity constraint and non-monotonic Lyapunov functions. In *Proceedings of the 5thInternational Conference on Control, Automation and Information Science*, 177–182, Ansan, South Korea.

163. Tri Tran, H.D. Tuan, Q.P. Ha, and H.T. Nguyen. 2011. Stabilising agent design for the control of interconnected systems. *International Journal of Control* 84 (6): 1140–1156.

164. Trinh, H., P.S. Teh, and T. Fernando. 2010. Time-delay systems: Design of delay-free and low-order observers. *IEEE Transactions on Automatic Control* 55 (10): 2434–2438.

165. Tuan, H.D., A. Savkin, T.N. Nguyen, and H.T. Nguyen. 2015. Decentralised model predictive control with stability constraints and its application in process control. *Journal of Process Control* 26: 73–89.

166. Van Der, A., and Schaft. 1996. L_2-Gain and Passivity Techniques in Nonlinear Control. Berlin: Springer.

167. Venkat, A.N., I.A. Hisken, J.B. Rawlings, and S.J. Wright. 2008. Distributed MPC strategies with applications to power system automatic generation control. *IEEE Transactions on Control Systems Technology* 16 (6): 1192–1206.

168. Vyatkin, V. 2013. Software engineering in industrial automation: State-of-the-art review. *IEEE Transactions on Industrial Informatics* 9 (3): 1234–1249.

169. Wakaiki, M., K. Okano, and J.P. Hespanha. 2017. Stabilization of systems with asynchronous sensors and controllers. *Automatica* 81: 314–321.

170. Walsh, C.C., and L.G. Bushnell. 2002. Stability analysis of networked control systems I. *IEEE Transactions on Control Systems Technology* 10: 438–446.

171. Wang, X., and M.D. Lemmon. 2011. Event-triggering in distributed networked control systems. *IEEE Transactions on Automatic Control* 56 (3): 586–601.

172. Willems, J.C. 1972. Dissipative dynamical systems, Parts I and II. *Arch. Rational Mechanics Analysis* 45: 321–393.

173. Willems, J.C. 2007. Dissipative dynamical systems. *European Journal of Control* 13: 134–151.

174. Wong, W.S., and R.W. Brockett. 1999. Systems with finite communication bandwidth constraints II. *IEEE Transactions on Automatic Control* 44: 1049–1053.

175. Wood, A.J., B.F. Woolenberg, and G.B. Sheblé. 2013. *Power Generation Operation and Control*, 3rd ed. New Jersey: Wiley.

176. Wu, M., Y. He, and J.-H. She. 2010. *Stability Analysis and Robust Control of Time-Delay Systems*. Berlin: Springer.

177. Xu, H., S. Jagannathan, and F.L. Lewis. 2012. Stochastic optimal control of unknown linear networked control system in the presence of random delays and packet losses. *Automatica* 48 (6): 1017–1030.

178. Yan, J., and V. Vyatkin. 2013. Distributed software architecture enabling peer-to-peer communicating controllers. *IEEE Transactions on Industrial Informatics* 9 (4): 2200–2209.

179. Yu, Q., and B. Wu. 2015. Generalized Lyapunov function theorems and its applications in switched systems. *Systems and Control Letters* 77: 40–45.

180. Zames, G. 1966. On the input-output stability of time-varying nonlinear feedback systems part I: Condition derived using concepts of loop gain, conicity, and positivity. *IEEE Transactions on Automatic Control* 11 (2): 228–238.

181. Zhang, J., Y. Lin, and P. Shi. 2015. Output tracking control of networked control systems via delay compensation controllers. *Automatica* 57 (6): 85–92.

Printed in the United States
By Bookmasters